T0233806

Lecture Notes in Computer Science 9211

Commenced Publication in 1973
Founding and Former Series Editors:
Gerhard Goos, Juris Hartmanis, and Jan van Leeuwen

Editorial Board

David Hutchison
Lancaster University, Lancaster, UK

Takeo Kanade
Carnegie Mellon University, Pittsburgh, PA, USA

Josef Kittler
University of Surrey, Guildford, UK

Jon M. Kleinberg
Cornell University, Ithaca, NY, USA

Friedemann Mattern
ETH Zurich, Zürich, Switzerland

John C. Mitchell
Stanford University, Stanford, CA, USA

Moni Naor
Weizmann Institute of Science, Rehovot, Israel

C. Pandu Rangan
Indian Institute of Technology, Madras, India

Bernhard Steffen
TU Dortmund University, Dortmund, Germany

Demetri Terzopoulos
University of California, Los Angeles, CA, USA

Doug Tygar
University of California, Berkeley, CA, USA

Gerhard Weikum
Max Planck Institute for Informatics, Saarbrücken, Germany

More information about this series at http://www.springer.com/series/7407

Andrew Phillips · Peng Yin (Eds.)

DNA Computing and Molecular Programming

21st International Conference, DNA 21
Boston and Cambridge, MA, USA, August 17–21, 2015
Proceedings

 Springer

Editors
Andrew Phillips
Microsoft Research
Cambridge
UK

Peng Yin
Wyss Institute
Boston, MA
USA

ISSN 0302-9743   ISSN 1611-3349 (electronic)
Lecture Notes in Computer Science
ISBN 978-3-319-21998-1   ISBN 978-3-319-21999-8   (eBook)
DOI 10.1007/978-3-319-21999-8

Library of Congress Control Number: 2015944724

LNCS Sublibrary: SL1 – Theoretical Computer Science and General Issues

Springer Cham Heidelberg New York Dordrecht London
© Springer International Publishing Switzerland 2015
This work is subject to copyright. All rights are reserved by the Publisher, whether the whole or part of the material is concerned, specifically the rights of translation, reprinting, reuse of illustrations, recitation, broadcasting, reproduction on microfilms or in any other physical way, and transmission or information storage and retrieval, electronic adaptation, computer software, or by similar or dissimilar methodology now known or hereafter developed.
The use of general descriptive names, registered names, trademarks, service marks, etc. in this publication does not imply, even in the absence of a specific statement, that such names are exempt from the relevant protective laws and regulations and therefore free for general use.
The publisher, the authors and the editors are safe to assume that the advice and information in this book are believed to be true and accurate at the date of publication. Neither the publisher nor the authors or the editors give a warranty, express or implied, with respect to the material contained herein or for any errors or omissions that may have been made.

Printed on acid-free paper

Springer International Publishing AG Switzerland is part of Springer Science+Business Media
(www.springer.com)

Preface

This volume contains the papers presented at DNA 21: the 21st International Conference on DNA Computing and Molecular Programming. The conference was held at the Wyss Institute for Biologically Inspired Engineering, Harvard University, Massachusetts, USA, during August 17–21, 2015, and organized under the auspices of the International Society for Nanoscale Science, Computation and Engineering (ISNSCE). The DNA conference series aims to draw together mathematics, computer science, physics, chemistry, biology, and nanotechnology to address the analysis, design, and synthesis of information-based molecular systems.

Presentations were sought in all areas that relate to biomolecular computing, including, but not restricted to: algorithms and models for computation on biomolecular systems; computational processes in vitro and in vivo; molecular switches, gates, devices, and circuits; molecular folding and self-assembly of nanostructures; analysis and theoretical models of laboratory techniques; molecular motors and molecular robotics; studies of fault-tolerance and error correction; software tools for analysis, simulation, and design; synthetic biology and in vitro evolution; applications in engineering, physics, chemistry, biology, and medicine.

Authors who wished to present their work were asked to select one of two submission tracks: Track A (full paper) or Track B (one-page abstract with supplementary document). Track B is primarily for authors submitting experimental results who plan to submit to a journal rather than publish in the conference proceedings. We received 63 submissions for oral presentations: 26 submissions in Track A and 37 submissions in Track B. Each submission was reviewed by at least three reviewers, with an average of four reviewers per paper. The Program Committee accepted 13 papers in Track A and 15 papers in Track B. This volume contains the papers accepted for Track A.

We express our sincere appreciation to our invited speakers: George Church, Sharon Glotzer, Lila Kari, Paul Rothemund, Leslie Valiant, and Chris Voigt. We would also like to thank all of the authors who contributed papers to these proceedings, and who presented papers and posters during the conference. Last but not least, the editors would like to thank the members of the Program Committee and the additional invited reviewers for their hard work in reviewing the papers and providing constructive comments to authors.

August 2015

Andrew Phillips
Peng Yin

Organization

DNA 21 was organized at the Wyss Institute for Biologically Inspired Engineering, Harvard University, in cooperation with the International Society for Nanoscale Science, Computation and Engineering (ISNSCE).

Program Committee

Andrew Phillips (Co-chair)	Microsoft Research, UK
Peng Yin (Co-chair)	Harvard University, USA
Luca Cardelli	Microsoft Research, UK
Hendrik Dietz	Technical University of Munich, Germany
David Doty	California Institute of Technology, USA
Shawn Douglas	University of California, San Francisco, USA
Andrew Ellington	University of Texas at Austin, USA
Elisa Franco	University of California, Riverside and California Institute of Technology, USA
Cody Geary	Aarhus University, Denmark
Kurt Gothelf	Aarhus University, Denmark
Natasha Jonoska	University of South Florida, USA
Lila Kari	University of Western Ontario, Canada
Ken Komiya	Tokyo Institute of Technology, Japan
Marta Kwiatkowska	University of Oxford, UK
Tim Liedl	Ludwig Maximilian University of Munich, Germany
Chengde Mao	Purdue University, USA
Pekka Orponen	Aalto University, Finland
Jennifer Padilla	Boise State University, USA
Matthew Patitz	University of Arkansas, USA
Lulu Qian	California Institute of Technology, USA
John Reif	Duke University, USA
Alfonso Rodríguez-Patón	Universidad Politécnica de Madrid, Spain
Yannick Rondelez	University of Tokyo, Japan
Rebecca Schulman	Johns Hopkins University, USA
David Soloveichik	University of California, San Francisco
Darko Stefanovic	University of New Mexico, USA
Chris Thachuk	California Institute of Technology, USA
Andrew Turberfield	University of Oxford, UK
Erik Winfree	California Institute of Technology, USA
Damien Woods	California Institute of Technology, USA

Organizing Committee

William Shih (Co-chair)	Harvard University, USA
Peng Yin (Co-chair)	Harvard University, USA
Sungwook Woo	Harvard University, USA
Elena Chen	Harvard University, USA

Steering Committee

Natasha Jonoska (Chair)	University of South Florida, USA
Luca Cardelli	Microsoft Research, UK
Anne Condon	University of British Columbia, Canada
Masami Hagiya	University of Tokyo, Japan
Lila Kari	University of Western Ontario, Canada
Satoshi Kobayashi	University of Electro-Communications, Japan
Chengde Mao	Purdue University, USA
Satoshi Murata	Tohoku University, Japan
John Reif	Duke University, USA
Grzegorz Rozenberg	University of Leiden, The Netherlands
Nadrian Seeman	New York University, USA
Friedrich Simmel	Technical University of Munich, Germany
Andrew Turberfield	University of Oxford, UK
Hao Yan	Arizona State University, USA
Erik Winfree	California Institute of Technology, USA

Sponsoring Institutions

Wyss Institute for Biologically Inspired Engineering, Harvard University
Air Force Office of Scientific Research
Army Research Office
National Science Foundation
Office of Naval Research

Additional Reviewers

Angeli, David
Barbot, Benoit
Beliveau, Brian
Boemo, Michael
Ceska, Milan
Chen, Xi
Clamons, Samuel
Condon, Anne
Dai, Mingjie
Dannenberg, Frits
de Los Santos,
    Emmanuel Lorenzo
Ellis-Monaghan, Joanna
Enaganti, Srujan Kumar
Evans, Constantine
Gopalkrishnan, Nikhil

Gopinath, Ashwin
Hendricks, Jacob
Johnson, Robert F.
Karamichalis, Rallis
Karpenko, Daria
Kim, Jin-Woo
Kim, Jongmin
Kopecki, Steffen
Kulkarni, Manasi
Lakin, Matthew R.
Liu, Ninning
Manuch, Jan
McQuillan, Ian
Mohammed,
    Abdul Majeed
Murphy, Niall

Myhrvold, Cameron
Oishi, Kevin
Ong, Luvena
Paoletti, Nicola
Petersen, Rasmus
    Lerchedahl
Rogers, Trent
Sainz de Murieta, Iñaki
Schiefer, Nicholas
Seki, Shinnosuke
Srinivas, Niranjan
Subramanian, Hari
Summers, Scott
Zandron, Claudio

Contents

Dominance and T-Invariants for Petri Nets and Chemical Reaction Networks

Robert Brijder[1,2]([⊠])

[1] Hasselt University, Hasselt, Belgium
[2] Transnational University of Limburg, Hasselt, Belgium
robert.brijder@uhasselt.be

Abstract. Inspired by Anderson et al. [J. R. Soc. Interface, 2014] we study the long-term behavior of discrete chemical reaction networks (CRNs). In particular, using techniques from both Petri net theory and CRN theory, we provide a powerful sufficient condition for a structurally-bounded CRN to have the property that none of the non-terminal reactions can fire for all its recurrent configurations. We compare this result and its proof with a related result of Anderson et al. and show its consequences for the case of CRNs with deficiency one.

1 Introduction

Chemical reaction network (CRN) theory studies the behavior of chemical systems. Traditionally, the primary focus is on continuous CRNs, where mass action kinetics is assumed, see, e.g., [2,8–10]. In this setting a state is determined by the concentration of each species and the system evolves through ordinary differential equations. However, in scenarios where the number of molecules is small one needs to resort to discrete CRNs. In a discrete CRN a state (also called configuration) is determined by the counts of each species, and one often associates a probability to each reaction. In this paper we consider only discrete CRNs, and so, from now on, by CRN we will always mean a discrete CRN.

A CRN essentially consists of a finite set of reactions such as $A + B \rightarrow 2B$, which means that during this reaction one molecule of species A and one molecule of species B are consumed and as a result two molecules of species B are produced. We may depict a CRN as a graph, the reaction graph, where the vertices are the left-hand and right-hand sides of reactions and the edges are the reactions, see Fig. 1 for an example. We focus in this paper on the long-term behavior of CRNs for which the number of molecules cannot grow unboundedly. For such CRNs, called structurally-bounded CRNs, each configuration eventually reaches a configuration c such that c is reachable from any configuration c' reachable from c (i.e., we can always go back to c). Such configurations are called recurrent. The CRN N of Fig. 1 is structurally-bounded.

Now, let us consider the CRN N' obtained from N by replacing every vertex by one molecule of a distinct species X_i, see Fig. 2. We easily observe that for N', the recurrent configurations are exactly those without molecules of species

© Springer International Publishing Switzerland 2015
A. Phillips and P. Yin (Eds.): DNA 2015, LNCS 9211, pp. 1–15, 2015.
DOI: 10.1007/978-3-319-21999-8_1

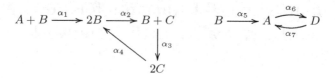

Fig. 1. The reaction graph of a CRN N.

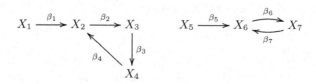

Fig. 2. The reaction graph of the CRN N' obtained from N by introducing a distinct species X_i for each vertex.

X_1 or X_5. In other words, the reactions β_1 and β_5 cannot fire for any recurrent configuration of N'. Notice that the reaction graph of N' has two strongly-connected components without outgoing edges: one having the vertices X_2, X_3, and X_4 and one having the vertices X_6 and X_7. The reactions outside these two strongly-connected components are called non-terminal. Thus N' has the property that none of the non-terminal reactions can fire for all its recurrent configurations. But what about the original CRN N? The dynamics of N are clearly more involved since we can go, for example, from configuration $A + B$ back to $A + B$ by firing reaction α_1 followed by firing reaction α_5.

The main result of this paper, cf. Theorem 1, is a sufficient condition for a structurally-bounded CRN to have the property that none of the non-terminal reactions can fire for all its recurrent configurations (we recall the notion of non-terminal reaction in Sect. 3). Those CRNs have relatively simple long-term behavior. The sufficient condition of Theorem 1 (when formulated in terms of so-called T-invariants in Corollary 2) is structural/syntactical and can be checked for many CRNs in a computationally-efficient way. Various non-trivial CRNs from the literature satisfy the sufficient condition of Theorem 1 (see, e.g., the CRNs given in [1]), and so it can make non-trivial predictions about the long-term behavior of those CRNs. In particular, the CRN N of Fig. 1 satisfies the sufficient condition. Moreover, this result can also be used as a tool for engineering CRNs that perform deterministic computations (independent of the probabilities), such as in the computational model of [5]. Indeed, such CRNs generally require relatively simple long-term behavior which may be partially verified by Theorem 1.

Theorem 1 is inspired by the main technical result of [1] (which in turn was inspired by the main result of [16]), which provides another sufficient condition for the non-applicability of non-terminal reactions for recurrent configurations. However, there are a number of differences between both results. First, Theorem 1 is derived in a basic combinatorial setting using notions from Petri net theory such as the notion of T-invariant, without considering stochastics. In contrast, the intricate proof of the main result of [1] is derived in a very different setting that uses non-trivial arguments from both mass action kinetics and stochastics. Secondly, we show examples where the main result of [1] is silent, while Theorem 1 makes a prediction. In fact, we conjecture that the main result of [1] is a special case of Theorem 1. We compare both results in detail in Sect. 4. While we focus in this paper on recurrent configurations of CRNs, we mention that the related concept of recurrent CRN has been investigated in [14].

While formulated in terms of CRNs, the results in this paper equally apply to Petri nets, which is a very well studied model of parallel computation, see, e.g., [15]. Using the "dictionary" provided for the reader with a Petri net background (see Subsect. 2.2), it is straightforward to reformulate the results in this paper in terms of Petri nets.

Due to space constraints, proofs of the results are omitted and a corollary concerning CRNs of deficiency one is omitted. They can be found in the full version of this paper [4].

2   Standard Graph and CRN/Petri Net Notions

2.1   Preliminaries

Let $\mathbb{N} = \{0, 1, \ldots\}$. Let X and Y be arbitrary sets. The set of vectors indexed by X with entries in Y (i.e., the set of functions $\varphi : X \to Y$) is denoted by Y^X. For $v, w \in \mathbb{N}^X$, we write $v \leq w$ if $v(x) \leq w(x)$ for all $x \in X$. Moreover, we write $v < w$ if $v \leq w$ and $v \neq w$. The *support* of v, denoted by $\mathrm{supp}(v)$, is the set $\{x \in X \mid v(x) > 0\}$. For finite sets X and Y, a $X \times Y$ matrix A is a matrix where the rows and columns are indexed by X and Y, respectively.

We consider digraphs $G = (V, E, F)$ where V and E are finite sets of vertices and edges and $F : E \to V^2$ assigns to each edge $e \in E$ an ordered vertex pair (u, v). We denote V by $V(G)$ and E by $E(G)$. The *incidence matrix* of G is the $V(G) \times E(G)$ matrix A where for $e \in E$ with $F(e) = (v, w)$ we have entries $A(v, e) = -1$, $A(w, e) = 1$, and $A(u, e) = 0$ for all $u \in V \setminus \{v, w\}$ if $v \neq w$, and $A(u, e) = 0$ for all $u \in V$ if $v = w$. The number of connected components of a digraph G is denoted by $c(G)$. It is well known that the rank $r(A)$ of the incidence matrix A of a digraph G is equal to $|V| - c(G)$ (where it does not matter over which field the rank is computed [13, Proposition 5.1.2]). From now on we let the field $\mathbb{Q}$ of rational numbers be the field in which we compute.

A walk π in G is described by (particular) strings over E. Let $\Phi(\pi)$ denote the *Parikh image* of π, i.e., $\Phi(\pi) \in \mathbb{N}^E$ where $(\Phi(\tau))(e)$ is the number of occurrences of e in π. We write $\mathrm{supp}(\pi) = \mathrm{supp}(\Phi(\pi))$, i.e., $\mathrm{supp}(\pi)$ is the set of elements that occur in π. The vectors v of $\ker(A) \cap \mathbb{N}^E$ describe the cycles of G, i.e., they describe the Parikh images of closed walks in G.

For convenience we identify a digraph G with its $V(G) \times E(G)$ incidence matrix. Hence, we may for example speak of the rank $r(G)$ of G. We say that $e \in E(G)$ is a *bridge* if e is not contained in any closed walk of G. The *induced subgraph* G' *of* G with respect to $X \subseteq V(G)$ is the digraph $G' = (X, E', F')$ where E' is the preimage of X^2 under F and F' is the restriction of F to E'. A *strongly connected component* (*SCC*, for short) is an induced subgraph G' of G with respect to $X \subseteq V(G)$ such that G' contains no bridge and X is largest (with respect to inclusion) with this property.

2.2   CRNs and Petri Nets

We now recall the notion of a chemical reaction network.

Definition 1. *A* chemical reaction network *(or* CRN *for short)* N *is a 3-tuple* (S, R, F) *where* S *and* R *are finite sets and* F *is a function that assigns to each* $r \in R$ *an ordered pair* $F(r) = (v, w)$ *where* $v, w \in \mathbb{N}^S$. *Vector* v *is denoted by* $\mathrm{in}(r)$ *and* w *by* $\mathrm{out}(r)$.

The elements of S are called the *species* of N, the elements of R are called the *reactions* of N, and F is called the *reaction function*. For a reaction r, $\mathrm{in}(r)$ and $\mathrm{out}(r)$ are called the *reactant vector* and *product vector* of r, respectively.

It is common in the literature of CRNs to omit the function F and have R as a set of tuples (v, w). However, this would not allow two different reactions to have the same reactant and product vectors (such situations are common in Petri net theory).

In CRN theory, it is common to write vectors in additive notation, so, e.g., if $S = \{A, B, C\}$, then $A + 2B$ denotes the vector v with $v(A) = 1$, $v(B) = 2$, and $v(C) = 0$.

Example 1. Consider the CRN $N = (S, R, F)$ with $S = \{A, B\}$, $R = \{a, b\}$, $F(a) = (A + B, 2B)$ and $F(b) = (B, A)$. This CRN is taken from [16] (see also [1]). This example is the running example of this section.

We now define a natural digraph for a CRN N, called the reaction graph of N. The name is from [11], and the concept is originally defined in [8].

Definition 2. *Let* $N = (S, R, F)$ *be a CRN. The* reaction graph *of* N, *denoted by* $\mathcal{R}_N$, *is the labeled digraph* (V, R, F) *with* $V = \{\mathrm{in}(r) \mid r \in R\} \cup \{\mathrm{out}(r) \mid r \in R\}$.

Note that in the reaction graph each reactant and product vector becomes a single vertex. The vertices of the reaction graph are called *complexes*. The reaction graph of the CRN N of our running example (Example 1) is depicted in Fig. 3.

$$A + B \xrightarrow{\ a\ } 2B \qquad B \xrightarrow{\ b\ } A$$

Fig. 3. The reaction graph of the CRN of Example 1.

A *configuration* c of N is a vector $c \in \mathbb{N}^S$. Let $r \in R$. We say that r can *fire* on c if $\mathrm{in}(r) \leq c$. In this case we also write $c \to^r c'$ where $c' = c - \mathrm{in}(r) + \mathrm{out}(r)$.

Note that c' is a configuration as well. Moreover, we write $c \to c'$ if $c \to^r c'$ for some $r \in R$. For $\tau \in R^*$ (as usual, R^* is Kleene star on R) we write $c \to^\tau c'$ if $c \to^{\tau_1} c_1 \cdots \to^{\tau_n} c'$ where $\tau = \tau_1 \cdots \tau_n$ and $\tau_i \in R$ for all $i \in \{1, \ldots, n\}$. The reflexive and transitive closure of the relation $\to$ is denoted by $\to^*$. If $c \to^* c'$, then we say that c' is *reachable* from c. We say that a configuration c is *recurrent* if for all c' with $c \to^* c'$ we have $c' \to^* c$. Note that if c is recurrent and $c \to^* c'$, then c' is recurrent.

Example 2. Consider again the running example. We have, e.g., $2A + B \to^{aabb} 2A + B$. However, $2A + B$ is not recurrent as $2A + B \to^b 3A$ and in configuration $3A$ no reaction can fire. In fact, the recurrent configurations of N are precisely those that do not contain any B. Indeed, assume c is recurrent. Then we can fire b until we obtain a configuration c' that does not contain any B. No reaction can fire for c' and so $c = c'$ since c is recurrent.

The definition of a CRN is equivalent to that of a Petri net [15]. In a Petri Net, species are called *places* p, reactions are called *transitions*, and configurations are called *markings*. A Petri net is often depicted as a graph with two types of vertices, one type for the places and one for the transitions. The Petri net-style depiction of the running example is given in Fig. 4. The round vertices are the places and the rectangular vertices are the transitions. We use in this paper several standard Petri net notions, which are recalled in the next subsection.

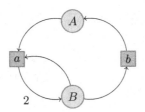

Fig. 4. The Petri net-style depiction of the running example.

2.3 P/T-Invariants

The notions of this subsection are all taken from Petri net theory [15]. We first recall the notion of an incidence matrix of a CRN, which is not to be confused with the notion of an incidence matrix of a digraph (as recalled above). In fact, we will compare in the next subsection the incidence matrix of a CRN with the incidence matrix of its reaction graph.

Definition 3. *For a CRN $N = (S, R, F)$, the incidence matrix of N, denoted by $\mathcal{I}_N$, is the $S \times R$ matrix A where for each $r \in R$ the column of A belonging to r is equal to $\mathrm{out}(r) - \mathrm{in}(r)$.*

Example 3. Consider again the CRN N of the running example. Then

$$\mathcal{I}_N = \begin{array}{c} \\ A \\ B \end{array}\begin{array}{c} a \quad b \\ \begin{pmatrix} -1 & 1 \\ 1 & -1 \end{pmatrix} \end{array}.$$

Note that if $c \to^\tau c'$, then $c' = c + \mathcal{I}_N \Phi(\tau)$, where $\Phi(\tau)$ denotes again the Parikh image of τ.

A $v \in \mathbb{N}^S$ is called a *P-invariant* of N if $v^T \mathcal{I}_N = 0$ (here 0 denotes a zero vector of suitable dimension indexed by R). Similarly, $v \in \mathbb{N}^R$ is called a *T-invariant* of N if $\mathcal{I}_N v = 0$, i.e., $v \in \ker(\mathcal{I}_N)$.[1] A P-invariant or T-invariant are also sometimes called P-semiflow and T-semiflow, respectively, in the literature. Observe that if $c \to^\tau c'$, then $\Phi(\tau)$ is a T-invariant if and only if $c' = c$. A CRN N is called *conservative* if there is a P-invariant v such that $\mathrm{supp}(v) = S$. Also, N is called *consistent* if there is a T-invariant v such that $\mathrm{supp}(v) = R$.

A CRN N is said to be *structurally bounded* when for every configuration c, there is a $k_c \in \mathbb{N}$ such that for each configuration c' with $c \to^* c'$ we have that each entry of c' is at most k_c. Note that for a structurally-bounded CRN, the number of different configurations reachable from a given configuration is finite, and so for each configuration c, there is a recurrent configuration reachable from c. In this way, one often informally views the recurrent configurations as the possible states of the CRN in "the long term".

The following result is well known.

Proposition 1 ([12]). *Let N be a CRN. If N is conservative, then N is structurally bounded.*

Example 4. The CRN N of the running example is both conservative and consistent. Indeed, any $v \in \mathbb{N}^S$ with $v(A) = v(B) \geq 1$ is a P-invariant with $\mathrm{supp}(v) = S$ and any $w \in \mathbb{N}^R$ with $v(a) = v(b) \geq 1$ is T-invariant with $\mathrm{supp}(v) = R$.

2.4   Deficiency

The notions that we recall in this subsection are originally from chemical reaction theory (and are less studied within Petri net theory).

Let $N = (S, R, F)$ be a CRN and let $V = \{\mathrm{in}(r) \mid r \in R\} \cup \{\mathrm{out}(r) \mid r \in R\}$. We denote by $\mathcal{Y}_N$ the $S \times V$ matrix with for all $s \in S$ and $v \in V$, entry $\mathcal{Y}_N(s, v)$ is equal to $v(s)$.

The next lemma relates the incidence matrix $\mathcal{I}_N$ of a CRN N with the incidence matrix of the reaction graph $\mathcal{R}_N$ of N.

Lemma 1 (Sect. 6 of [9]). *Let $N = (S, R, F)$ be a CRN. Then $\mathcal{I}_N = \mathcal{Y}_N \mathcal{R}_N$.*

---

[1] The P and T in P/T-invariant are short for Place and Transition (from Petri net theory). We choose to use these well-known names instead of calling them "S-invariant" and "R-invariant" for Species and Reaction, respectively.

In the above equality, $\mathcal{R}_N$ denotes the incidence matrix $\mathcal{R}_N$ and not the graph.

As a corollary to Lemma 1, we have the following.

Corollary 1 ([11]). *Let $N = (S, R, F)$ be a CRN. Then* $\ker(\mathcal{R}_N) \subseteq \ker(\mathcal{I}_N)$.

The vectors v of $\ker(\mathcal{R}_N) \cap \mathbb{N}^R$, which are T-invariants by Corollary 1, are called *closed* T-invariants [3]. Recall that the vectors v of $\ker(\mathcal{R}_N) \cap \mathbb{N}^R$ describe the cycles of $\mathcal{R}_N$, and so for each closed T-invariant v of N, supp(v) does not contain any bridge of $\mathcal{R}_N$. Since each of the entries of a T-invariant is nonnegative, the linear space $\ker(\mathcal{I}_N)$ does not necessarily have a basis consisting of only T-invariants, see Example 5 below.

The *deficiency* $\delta(N)$ of a CRN N is $r(\mathcal{R}_N) - r(\mathcal{I}_N)$. By Corollary 1, $\delta(N)$ is non-negative. Thus, one may view $\delta(N)$ as a measure of the difference in dimensions between $\ker(\mathcal{R}_N)$ and $\ker(\mathcal{I}_N)$. The former is determined only by the structure of the reaction graph (ignoring the identity of the vertices), while the latter also incorporates the relations that rely on the identities of the vertices of the reaction graph.

Recall from Subsect. 2.1 that $r(\mathcal{R}_N) = |V(\mathcal{R}_N)| - c(\mathcal{R}_N)$. Hence, we have $\delta(N) = |V(\mathcal{R}_N)| - c(\mathcal{R}_N) - r(\mathcal{I}_N)$ [8,10]. Note that if $\delta(N) = 0$, then every T-invariant of N is closed and $\ker(\mathcal{R}_N) = \ker(\mathcal{I}_N)$.

$$A + B \xrightarrow{\ a\ } 2B \qquad A \xrightarrow{\ b\ } B$$

Fig. 5. The reaction graph of a CRN discussed in Example 5.

Example 5. In the running example, $\ker(\mathcal{R}_N)$ only contains the zero vector, while $\ker(\mathcal{I}_N)$ contains all scalar multiples of the vector w with $w(a) = w(b) = 1$. Thus $\ker(\mathcal{I}_N)$ has a basis consisting of only T-invariants. Moreover, $\delta(N) = 1$. Alternatively, the reaction graph $\mathcal{R}_N$ has 4 vertices and 2 connected components and $r(\mathcal{I}_N) = 1$. Thus, $\delta(N) = 4 - 2 - 1 = 1$.

If we consider the CRN N' of Fig. 5, then $\ker(\mathcal{R}_{N'})$ also only contains the zero vector, while $\ker(\mathcal{I}_{N'})$ contains all scalar multiples of the vector w with $w(a) = -w(b) = 1$. Again, $\delta(N') = 1$, however the only T-invariant of $\ker(\mathcal{I}_{N'})$ is the zero vector.

3 Dominance and Non-closed T-Invariants

Note that there is a natural partial order for the set of SCCs of a graph: for SCCs X and Y, we have $X \preccurlyeq Y$ if there is a path from a vertex of Y to a vertex of X. We now consider a different partial order, denoted by $\leq_d$, for the SCCs of a reaction graph of a CRN.

Let N be a CRN. For SCCs X and Y of $\mathcal{R}_N$ we write $X \leq_d Y$ if there are vertices x of X and y of Y such that $x \leq y$.

Lemma 2. *Let $N = (S, R, F)$ be a structurally-bounded CRN. Then the $\leq_d$ relation between SCCs of $\mathcal{R}_N$ is a partial order.*

For SCCs X and Y we write $X <_d Y$ if $X \leq_d Y$ and $X \neq Y$. We say that X *dominates* Y when $X <_d Y$. For a set $\mathcal{S}$ of SCCs, we let $\min_{\leq_d}(\mathcal{S}) \subseteq \mathcal{S}$ be the set of elements of $\mathcal{S}$ that are minimal with respect to the $\leq_d$ relation among all the elements of $\mathcal{S}$.

Let us define for a SCC X of $\mathcal{R}_N$, $\text{out}(X) = \{r \in E(\mathcal{R}_N) \mid \text{in}(r) \in V(X), \text{out}(r) \notin V(X)\}$. We call X *terminal* if $\text{out}(X) = \emptyset$. We call a reaction r (complex x, resp.) *terminal* if $r \in E(X)$ ($x \in V(X)$, resp.) for some terminal SCC X of $\mathcal{R}_N$.

We will consider the minimal set $\mathcal{X}$ of non-terminal SCCs that dominates all other non-terminal SCCs. In other words, if we let $\mathcal{N}$ be the set of non-terminal SCCs, then $\mathcal{X} = \min_{\leq_d}(\mathcal{N})$.

Let B be the set of bridges of $\mathcal{R}_N$. The *exit set* of a set $\mathcal{S}$ of non-terminal SCCs, is a set $Z \subseteq B$ with both $|Z| = |\mathcal{S}|$ and $|Z \cap \text{out}(X)| = 1$ for all $X \in \mathcal{S}$. In other words, Z contains exactly one bridge of $\text{out}(X)$ for each $X \in \mathcal{S}$.

Assuming the existence of a non-terminal reaction that can fire for some recurrent configuration c, the main result of this paper ensures the existence of certain sequences τ with $c' \to^\tau c'$ for some configuration c' reachable from c. For each exit set Z, there exists such a τ that avoids all bridges outside Z and, at the same time, uses the bridges of Z whenever possible. As a consequence, each of the sequences τ corresponds to a T-invariant $v = \Phi(\tau)$ that have zero entries for the bridges outside Z and nonzero entries for some of the bridges inside Z. We will show that for various CRNs this necessary condition allows one to show that only terminal reactions can fire for all its recurrent configurations.

The proof idea is the following. Let us start with a recurrent configuration c. While traversing the configuration space by applying reactions starting from c, we need never choose a bridge of $\mathcal{R}_N$ going out of a SCC X that is dominated by some Y (i.e., $Y <_d X$). Indeed, if $x \in V(X)$ and $y \in V(Y)$ with $y < x$, then we may walk inside X to x and $y < x$ implies that any reaction r with $\text{in}(r) = y$ can fire for x. In this way we also avoid taking a reaction r' with $\text{in}(r') = x$. Moreover, walking out of Y can be done by taking any of the bridges. We choose the one from the exit set Z. Now, eventually, our path inside $\mathcal{R}_N$ will lead to a terminal vertex. However, since c is recurrent, we can go back to c. If a non-terminal reaction can fire for c, then this means that we can iterate this process (walking along bridges, etc.). Structural boundedness finally ensures that the configuration space is finite and so, we must eventually repeat a configuration that closes the "circuit".

We are now ready to formulate the main result of this paper.

Theorem 1. *Let $N = (S, R, F)$ be a structurally-bounded CRN, and let $\mathcal{X} = \min_{\leq_d}(\mathcal{N})$, where $\mathcal{N}$ is the set of non-terminal SCCs of $\mathcal{R}_N$. Let B be the set of bridges of $\mathcal{R}_N$. Let L be the set of all non-terminal reactions r of $\mathcal{R}_N$ such that there is a non-terminal reaction r' of $\mathcal{R}_N$ with $\text{in}(r') < \text{in}(r)$.*

If some non-terminal reaction can fire for some recurrent configuration c, then for each exit set Z of $\mathcal{X}$, there is a $\tau \in R^*$ such that

1. τ contains no reactions from $(B \setminus Z) \cup L$,
2. $\tau = \pi_1\sigma_1 \cdots \pi_n\sigma_n$ where each π_i is a path in $\mathcal{R}_N$ from a non-terminal vertex to a terminal vertex and each σ_i is a sequence of terminal reactions, and
3. $c' \to^\tau c'$ for some recurrent configuration c' reachable from c.

$$A \xrightarrow{a} E \qquad C \xrightarrow{b} D \qquad E + D \xrightarrow{d} A + C$$

Fig. 6. The reaction graph of the CRN of Example 6.

We illustrate Theorem 1 through a couple of examples.

Example 6. Consider the CRN N of Fig. 6. It is easy to verify that $c = A + C$ is a recurrent configuration. Moreover, there is a non-terminal reaction r that can fire for this configuration (take $r = a$ or $r = b$). Note that there is only one exit set Z for $\mathcal{X}$, which is $Z = B = \{a, b, d\}$. By Theorem 1, there is a $\tau \in R^*$ such that (1) τ contains no reactions from $(B \setminus Z) \cup L$, (2) τ is a sequence of paths, each going to a terminal vertex, and (3) $c' \to^\tau c'$ for some recurrent configuration c' reachable from c. Indeed, we can choose, e.g., $\tau = abd$ and $c' = A + C$.

$$A + C \xrightarrow{a} E + C \qquad E + D \xrightarrow{b} A + D$$

Fig. 7. The reaction graph of the CRN of Example 7.

We now give another example.

Example 7. Consider the CRN N of Fig. 7. It is easy to verify that $c = A + C + D$ is a recurrent configuration. Moreover, there is a non-terminal reaction r that can fire for this configuration (take $r = a$). We have that $Z = B = \{a, b\}$ is the unique exit set Z for $\mathcal{X}$. We notice that $\tau = ab$ and $c' = A + C + D$ satisfy the conditions of Theorem 1. Indeed, we have $\tau = \tau_1\sigma_1\tau_2\sigma_2$ with $\tau_1 = a$, $\sigma_1 = \epsilon$ (the empty string), $\tau_2 = b$, and $\sigma_2 = \epsilon$. Note that if N contained the additional reaction $A + D \to^d E + D$, then $\tau = ab$ and $c' = A + C + D$ would again satisfy the conditions of Theorem 1, where $\tau = \tau_1\sigma_1$ with $\tau_1 = a$ and $\sigma_1 = b$.

Considering the non-closed T-invariant $v = \Phi(\tau)$ with τ from Theorem 1, we have the following corollary to Theorem 1. Note that Condition 2 of Theorem 1 implies that $\text{supp}(v)$ contains a bridge, and therefore $v(z) \neq 0$ for some $z \in Z$.

Corollary 2. *Let N, $\mathcal{X}$, B, and L be as in Theorem 1.*

Assume there is an exit set Z of $\mathcal{X}$ such that there is no non-closed T-invariant v with (1) $v(x) = 0$ for all $x \in (B \setminus Z) \cup L$ and (2) $v(z) \neq 0$ for some $z \in Z$.

Then no non-terminal reaction can fire for any recurrent configuration of N.

We remark that, in view of Theorem 1, Corollary 2 can be strengthened by replacing the condition $v(z) \neq 0$ for some $z \in Z$ with the stronger (but more involved) condition that says that the (occurrences of the) non-terminal reactions of v form a set of paths where each path ends in a terminal vertex.

Note that since closed T-invariants v cannot contain bridges, we may without loss of generality remove the condition that v is "non-closed" in Corollary 2.

We use Corollary 2 to determine whether no non-terminal reaction can fire for any recurrent configuration of a CRN. While non-closed T-invariants have a central role in Corollary 2, curiously, this notion from [3] has been given only modest attention in both the Petri net theory and the CRN theory.

For a given exit set Z of $\mathcal{X}$, one can verify using linear programming in polynomial time whether or not there is a non-closed T-invariant v with the properties of Corollary 2. While in general there may be an exponential number of exit sets (exponential in the number of reactions) to check, in many cases the number of exit sets is severely constraint and in these cases the sufficient condition of Corollary 2 is computationally efficient.

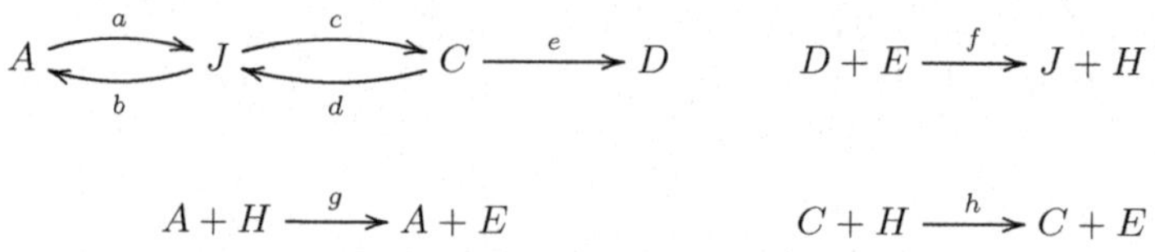

$$
A \overset{a}{\underset{b}{\rightleftharpoons}} J \overset{c}{\underset{d}{\rightleftharpoons}} C \overset{e}{\longrightarrow} D \qquad\qquad D + E \overset{f}{\longrightarrow} J + H
$$

$$
A + H \overset{g}{\longrightarrow} A + E \qquad\qquad\qquad C + H \overset{h}{\longrightarrow} C + E
$$

Fig. 8. The reaction graph of the CRN of Example 8.

We now give some examples to illustrate Corollary 2.

Example 8. Consider the CRN N of Fig. 8. This CRN is a simplification of a CRN from biology studied in [16] (see also [1]). We have

$$
\mathcal{I}_N =
\begin{array}{c} \\ A \\ J \\ C \\ D \\ E \\ H \end{array}
\begin{array}{c}
\begin{array}{cccccccc} a & b & c & d & e & f & g & h \end{array} \\
\left(
\begin{array}{cccccccc}
-1 & 1 & 0 & 0 & 0 & 0 & 0 & 0 \\
1 & -1 & -1 & 1 & 0 & 1 & 0 & 0 \\
0 & 0 & 1 & -1 & -1 & 0 & 0 & 0 \\
0 & 0 & 0 & 0 & 1 & -1 & 0 & 0 \\
0 & 0 & 0 & 0 & 0 & -1 & 1 & 1 \\
0 & 0 & 0 & 0 & 0 & 1 & -1 & -1
\end{array}
\right)
\end{array}.
$$

It is easy to verify that the sum of the rows of $\mathcal{I}_N$ is the zero vector and so N is conservative. Consequently, N is structurally bounded. It turns out that $\ker(\mathcal{I}_N)$ is of dimension 4 and is spanned by T-invariants. In fact, one can verify that $\ker(\mathcal{I}_N)$ is spanned by the two closed T-invariants $w_1 = \Phi(ab)$ and $w_2 = \Phi(cd)$

together with the two non-closed T-invariants $v_1 = \Phi(gfce)$ and $v_2 = \Phi(hfce)$. We remark that $A+H+D \to^{gfce} A+H+D$ and $C+H+D \to^{hfce} C+H+D$. Thus $\delta(N) = 2$. Note that $B = \{e, f, g, h\}$ is the set of bridges of $\mathcal{R}_N$. Let $\mathcal{X}$ be the set of non-terminal SCCs of $\mathcal{R}_N$ that are minimal with respect to $\leq_d$. We notice that $Z = \{e, f\}$ is the only exit set of $\mathcal{X}$. Also $L = \{g, h\}$. Now, the non-closed T-invariants v_1 and v_2 are witnesses that there is no non-closed T-invariant v with both (1) $v(g) = v(h) = 0$ (note that $(B \setminus Z) \cup L = \{g, h\}$) and (2) either $v(e)$ or $v(f)$ nonzero. By Corollary 2, for every recurrent configuration no non-terminal reaction can fire. Since every reaction is non-terminal, for every recurrent configuration no reaction can fire.

$$A + B \xrightarrow{a} 2B \qquad B + C \xrightarrow{b} A + C$$

Fig. 9. The reaction graph of the CRN of Example 9.

The next example shows that the converse of Theorem 1 does not hold.

Example 9. Consider the CRN N of Fig. 9. We show that no reaction can fire for any recurrent configuration of N. Let c be a recurrent configuration. If c does not contain any C, then we can fire reaction a until we obtain a configuration c' for which no more reactions can fire. Since c is recurrent, $c = c'$ and we are done. If c contains at least one C, then we can apply reaction b until we obtain a configuration c'' with only A's and C's. Hence no reaction can fire for c''. Since c is recurrent, we have $c = c''$ and we are done.

However, for $c = A + B + C$ we have $c \to^\tau c$ with $\tau = ab$. We notice that $Z = \{a, b\}$ is the only exit set of $\mathcal{X}$ and $(B \setminus Z) \cup L = \emptyset$. Thus τ trivially contains no reactions from $(B \setminus Z) \cup L$ and $\tau = \pi_1 \pi_2$ with paths $\pi_1 = a$ and $\pi_2 = b$ in $\mathcal{R}_N$ from non-terminal vertices to terminal vertices. This shows that the converse of Theorem 1 does not hold.

We remark that if we remove species C from reaction b (in this way obtaining the running example of Sect. 2), then Corollary 2 (and Theorem 1) would have been applicable to show that no (non-terminal) reaction can fire for any recurrent configuration of N.

4   Using Rates

This paper is inspired by the main technical result of [1] (cf. Theorem 3.3 of the supplementary material of [1]). In this section we recall its result. First we recall a particular matrix. Let $\mathbb{R}_{\geq 0}$ ($\mathbb{R}_{>0}$, resp.) be the set of nonnegative (positive, resp.) real numbers.

Definition 4. *Let $N = (S, R, F)$ be a CRN. Let $V = V(\mathcal{R}_N)$ and let $\kappa \in \mathbb{R}_{>0}^R$. We denote by $\mathcal{K}_{N,\kappa}$ the $S \times V$ matrix where for each $x \in V$ the column of $\mathcal{K}_{N,\kappa}$ belonging to x is equal to $\sum_{r \in R, \text{in}(r)=x} \kappa(r) \cdot (\text{out}(r) - \text{in}(r))$.*

The value $\kappa(r)$ in Theorem 2 may be interpreted as the "rate" of reaction r. Note that the definition of $\mathcal{K}_{N,\kappa}$ is closely related to the definition of $\mathcal{I}_N$ (Definition 3).

We are now ready to formulate the main technical result of [1].

Theorem 2 ([1]). *Let $N = (S, R, F)$ be a conservative CRN and $V = V(\mathcal{R}_N)$. Let L be the set of non-terminal vertices v of $\mathcal{R}_N$ such that there is a non-terminal vertex v' of $\mathcal{R}_N$ with $v' < v$. Assume that $L \neq \emptyset$.*

If some non-terminal reaction can fire for some recurrent configuration c, then for all $\kappa \in \mathbb{R}_{>0}^R$, there is a $w \in \ker(\mathcal{K}_{N,\kappa}) \cap \mathbb{R}_{\geq 0}^V$ with $\mathrm{supp}(w) \cap L = \emptyset$ and there is a non-terminal vertex x with $x \in \mathrm{supp}(w)$.

Theorem 2 is proved in [1] using both intricate probabilistic arguments and methods from mass action kinetics. In [1], the theorem is unnecessarily stated in a probabilistic fashion using the notion of "positive recurrent configuration" for stochastically modeled CRNs: it can be stated in a deterministic way (see Theorem 2 above) by realizing that the configuration space is finite for a given initial configuration in a structurally-bounded CRN. This deterministic formulation and the discrete model (in contrast to mass action) triggered the search of this paper for a combinatorial explanation of this result. We invite the reader to compare the proof techniques used to prove Theorem 2 in [1] and Theorem 1 in this paper.

$$2B \xleftarrow{\;b\;} A + B \xrightarrow{\;a\;} 2A$$

Fig. 10. The reaction graph of the CRN of Example 10.

Note that if $L = \emptyset$, then Theorem 2 is silent. We now show an example with $L = \emptyset$ where Corollary 2 can be applied.

Example 10. Consider the CRN N of Fig. 10. Note that N is conservative with $w(A) = w(B) = 1$ as a witness. The only T-invariants v of N are those where $v(a) = v(b)$. Let $Z = \{a\}$ be an exit set of $\mathcal{X}$. Then there is no non-closed T-invariant v with $v(b) = 0$ and $v(a) \neq 0$. By Corollary 2, no non-terminal reaction can fire for any recurrent configuration c of N. Since all reactions of N are non-terminal, no reaction can fire for any recurrent configuration c of N. Indeed, one observes that the recurrent configurations of N are those configurations containing either only A's or only B's, for which a and b cannot fire.

We conjecture that the assumption $L \neq \emptyset$ can be removed from Theorem 2. In case $L \neq \emptyset$ is removed from Theorem 2, then Theorem 2 also predicts that no non-terminal reaction can fire for any recurrent configuration of the CRN of Example 10. Next, we give an example with $L \neq \emptyset$, where Corollary 2 can be applied but Theorem 2 is silent.

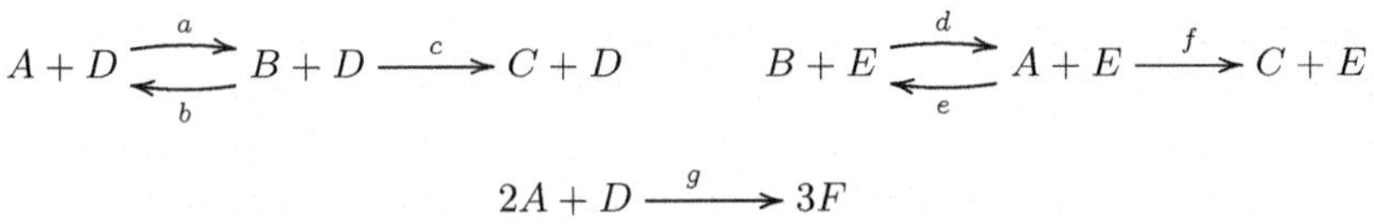

$$A + D \underset{b}{\overset{a}{\rightleftarrows}} B + D \overset{c}{\longrightarrow} C + D \qquad B + E \underset{e}{\overset{d}{\rightleftarrows}} A + E \overset{f}{\longrightarrow} C + E$$

$$2A + D \overset{g}{\longrightarrow} 3F$$

Fig. 11. The reaction graph of the CRN of Example 11.

Example 11. Consider the CRN N of Fig. 11. Note that N is conservative with $w(X) = 1$ for all species X as a witness. Note that $A + D < 2A + D$ and so $L \neq \emptyset$ in Theorem 2. Let $\kappa \in \mathbb{R}^R_{>0}$. We have $\mathcal{K}_{N,\kappa} =$

	$A+D$	$B+D$	$B+E$	$A+E$	$2A+D$	$C+D$	$C+E$	$3F$
A	$-\kappa(a)$	$\kappa(b)$	$\kappa(d)$	$-\kappa(e)-\kappa(f)$	$-2\kappa(g)$	0	0	0
B	$\kappa(a)$	$-\kappa(b)-\kappa(c)$	$-\kappa(d)$	$\kappa(e)$	0	0	0	0
C	0	$\kappa(c)$	0	$\kappa(f)$	0	0	0	0
D	0	0	0	0	$-\kappa(g)$	0	0	0
E	0	0	0	0	0	0	0	0
F	0	0	0	0	$3\kappa(g)$	0	0	0

Let $w \in \mathbb{R}^V_{\geq 0}$ with $\kappa(a)w(A+D) = \kappa(d)w(B+E) > 0$ and $w(x) = 0$ for all other $x \in V$. Then $w \in \ker(\mathcal{K}_{N,\kappa}) \cap \mathbb{R}^V_{\geq 0}$ with $x \in \mathrm{supp}(w)$ for some non-terminal vertex x and $\mathrm{supp}(w) \cap L = \emptyset$. Thus Theorem 2 is silent. On the other hand, none of the non-closed T-invariants of N contains a bridge and so by Corollary 2, no non-terminal reaction can fire for any recurrent configuration of N.

Conversely, despite trying numerous examples, we could not find an example where Theorem 2 predicts that no non-terminal reaction can fire for any recurrent configuration, but where Theorem 1 is silent.

5 Discussion

Based on structural properties of CRNs, the main result of this paper (cf. Theorem 1) provides a sufficient condition to analyze the long-term behavior of CRNs. While its proof is using basic combinatorial arguments, the result is powerful enough to apply to a large class of CRNs. Also, the sufficient condition is computationally-efficient to verify for many CRNs. Another such sufficient condition is shown in [1], cf. Theorem 2. We have shown examples of CRNs where Theorem 1 is applicable while Theorem 2 is silent.

Given that discrete CRNs are equivalent to Petri nets, it is curious that the corresponding research areas of CRN theory and Petri net theory have evolved almost independently. In this paper we shown that notions from Petri net theory (in particular, T-invariance) are useful for CRN theory. Similarly, notion such as deficiency, originating from CRN theory, are useful for Petri net theory. At the interface of these two notions is the scarcely-studied notion of non-closed

T-invariant, which is crucial in the sufficient condition of Corollary 2. This illustrates that both research areas can significantly profit from each other.

An open problem is resolving whether Theorem 2 is indeed a special case of Theorem 1. Another open problem is to somehow strengthen Theorem 1 in a natural way to make it applicable for CRNs such as the one presented in Example 9.

A further research direction is to incorporate probabilities. One may associate a probability to each T-invariant by multiplying the probabilities of the corresponding reactions. An open problem is to find a probabilistic version of Theorem 1 to make predictions about long-term behavior of probabilistic computational models of CRNs, such as the models of [6,7,17].

Acknowledgements. We thank David Anderson for kindly explaining his work during the Banff International Research Station (BIRS) workshop on CRNs (14w5167). Also, we thank the organizers of this workshop during which this research was initiated. We are indebted to Matthew Johnston for carefully reading an earlier version of this paper and for providing useful comments. And in particular for finding a counterexample to a conjecture in an earlier version of this paper. We finally thank the five referees for their useful comments. R.B. is a postdoctoral fellow of the Research Foundation – Flanders (FWO).

References

1. Anderson, D.F., Enciso, G.A., Johnston, M.D.: Stochastic analysis of biochemical reaction networks with absolute concentration robustness. J. R. Soc. Interface **11**(93), 20130943 (2014)
2. Aris, R.: Prolegomena to the rational analysis of systems of chemical reactions. Arch. Ration. Mech. Anal. **19**(2), 81–99 (1965)
3. Boucherie, R.J., Sereno, M.: On closed support T-invariants and the traffic equations. J. Appl. Probab. **35**(2), 473–481 (1998)
4. Brijder, R.: Dominance and deficiency for Petri nets and chemical reaction networks. arXiv preprint, arXiv:1503.04005 (2015)
5. Chen, H.-L., Doty, D., Soloveichik, D.: Deterministic function computation with chemical reaction networks. In: Stefanovic, D., Turberfield, A. (eds.) DNA 2012. LNCS, vol. 7433, pp. 25–42. Springer, Heidelberg (2012)
6. Cook, M., Soloveichik, D., Winfree, E., Bruck, J.: Programmability of chemical reaction networks. In: Condon, A., Harel, D., Kok, J.N., Salomaa, A., Winfree, E. (eds.) Algorithmic Bioprocesses. Natural Computing Series, pp. 543–584. Springer, Berlin Heidelberg (2009)
7. Cummings, R., Doty, D., Soloveichik, D.: Probability 1 computation with chemical reaction networks. In: Murata, S., Kobayashi, S. (eds.) DNA 2014. LNCS, vol. 8727, pp. 37–52. Springer, Heidelberg (2014)
8. Feinberg, M.: Complex balancing in general kinetic systems. Arch. Ration. Mech. Anal. **49**(3), 187–194 (1972)
9. Feinberg, M., Horn, F.: Chemical mechanism structure and the coincidence of the stoichiometric and kinetic subspaces. Arch. Ration. Mech. Anal. **66**(1), 83–97 (1977)

10. Horn, F.: Necessary and sufficient conditions for complex balancing in chemical kinetics. Arch. Ration. Mech. Anal. **49**(3), 172–186 (1972)
11. Mairesse, J., Nguyen, H.: Deficiency zero Petri nets and product form. Fundam. Inf. **105**(3), 237–261 (2010)
12. Memmi, G., Roucairol, G.: Linear algebra in net theory. In: Brauer, W. (ed.) Net Theory and Applications. Lecture Notes in Computer Science, vol. 84, pp. 213–223. Springer, Heidelberg (1975)
13. Oxley, J.: Matroid theory, 2nd edn. Oxford University Press, Oxford (2011)
14. Paulevé, L., Craciun, G., Koeppl, H.: Dynamical properties of discrete reaction networks. J. Math. Biol. **69**(1), 55–72 (2014)
15. Reisig, W., Rozenberg, G. (eds.): APN 1998. LNCS, vol. 1491. Springer, Heidelberg (1998)
16. Shinar, G., Feinberg, M.: Structural sources of robustness in biochemical reaction networks. Science **327**(5971), 1389–1391 (2010)
17. Soloveichik, D., Cook, M., Winfree, E., Bruck, J.: Computation with finite stochastic chemical reaction networks. Nat. Comput. **7**(4), 615–633 (2008)

Synthesizing and Tuning Chemical Reaction Networks with Specified Behaviours

Neil Dalchau, Niall Murphy$^{(\boxtimes)}$, Rasmus Petersen, and Boyan Yordanov

Microsoft Research, Cambridge, CB1 2FB, UK
{ndalchau,a-nimurp,a-rapete,yordanov}@microsoft.com

Abstract. We consider how to generate chemical reaction networks (CRNs) from functional specifications. We propose a two-stage approach that combines synthesis by satisfiability modulo theories and Markov chain Monte Carlo based optimisation. First, we identify candidate CRNs that have the *possibility* to produce correct computations for a given finite set of inputs. We then optimise the reaction rates of each CRN using a combination of stochastic search techniques applied to the chemical master equation, simultaneously improving the *probability* of correct behaviour and ruling out spurious solutions. In addition, we use techniques from continuous time Markov chain theory to study the expected termination time for each CRN. We illustrate our approach by identifying CRNs for majority decision-making and division computation, which includes the identification of both known and unknown networks.

Keywords: Chemical reaction networks · Program synthesis · Parameter optimisation · Chemical master equation · Satisfiability modulo theories · Markov chain Monte Carlo

1 Introduction

A central goal of molecular programming is to be able to implement arbitrary dynamical behaviours. Chemical reaction networks (CRNs) are a popular formalism for describing biochemical systems, such as protein interaction networks, gene regulatory networks, synthetic logic circuits and molecular programs built from DNA. Extensive theoretical understanding exists about the behaviour of a multitude of CRNs, and the behaviour of some networks has been exhaustively explored [1]. Besides describing chemical systems, CRNs provide a common language for expressing problems studied in computer science theory (e.g. Petri nets, network protocols) as well as control theory and engineering. Methods exist to convert CRNs into equivalent physical implementations, based on DNA strand displacement [2,3] the DNA toolbox system [4] and genelets [5]. Therefore, we sought to develop a methodology for proposing candidate CRNs that exhibit a pre-specified behaviour.

The computational power of CRNs has been extensively studied [6]. It is known that error-free (stably computing [7]) CRNs compute exactly the class of

© Springer International Publishing Switzerland 2015
A. Phillips and P. Yin (Eds.): DNA 2015, LNCS 9211, pp. 16–33, 2015.
DOI: 10.1007/978-3-319-21999-8_2

semi-linear functions [8,9]. However, if the stability restriction is relaxed and we allow the CRN to sometimes compute the wrong answer then it is possible to implement a register machine, that is, CRNs with error can compute functions beyond the semi-linear class (indeed they are equivalent in power to Turing machines) [6,10].

Although there are procedures to generate CRNs for semi-linear functions [8,10], primitive recursive functions [6], or even from arbitrary Turing machines [6], the proposal of practical (i.e. experimentally implementable) CRNs that compute a given function has thus far mostly been a manual effort. In this work, we attempt to automate the proposal of CRNs, by formally specifying a behaviour and automatically identifying CRNs that satisfy the desired behaviour with high probability. First, we formalise the problem of identifying CRNs that have the capacity to produce correct, finite computations for a given finite set of inputs. This corresponds to a synthesis problem, as opposed to verification, where the goal is to determine the correctness of a given CRN [11]. We express CRN synthesis as a satisfiability modulo theories (SMT) problem, which can be addressed using solvers such as Z3 [12]. This allows us to generate a number of candidate CRNs or to prove that no such CRN of a given size (in terms of numbers of reactions, species and computation lengths) exists. However, while the existence of correct computations is guaranteed for each generated CRN, the probability of these computations might be low.

To determine whether correct computations can occur with high probability, we next optimise the reaction rates of each generated CRN. To solve the optimisation problem, we combine stochastic search strategies based on Markov chain Monte Carlo (MCMC) with numerical integration of the chemical master equation (CME). This part of the problem was recently addressed in [13,14], though applied only to a single input.

In this paper, we specifically focus on uniform CRNs, those that have a fixed number of species and reactions for all input sizes. We also restrict our attention to bimolecular CRNs, where there are precisely 2 reactants and 2 products in every reaction. Bimolecular CRNs are equivalent to Population Protocols (PPs) [7] and also guarantee that mass is conserved in the system. We applied our two-step approach first to majority decision-making, in which the network seeks to identify which of two inputs is in an initial majority. Majority networks are well-studied in the literature, and there are many known CRNs that give approximate solutions [15–17]. We then applied our approach to division, a non-linear function which has been relatively less studied. We show a range of CRNs for majority and division identified automatically using our method, some of which have been identified and characterised previously, though some of which are entirely novel. This illustrates the potential for automatically determining CRNs with a specified behaviour.

2   Preliminaries

A chemical reaction network (CRN) is a tuple $\mathcal{C} = (\Lambda, \mathcal{R})$, where $\Lambda = \{s_0, \ldots, s_n\}$ and $\mathcal{R} = \{r_0, \ldots, r_m\}$ denote the finite sets of species and reactions, respectively.

A reaction is a tuple $r = (\mathbf{r}^r, \mathbf{p}^r, k^r)$ where $\mathbf{r}^r$ and $\mathbf{p}^r$ are the reactant and product *stoichiometry* vectors ($\mathbf{r}^r_s \in \mathbb{N}_0$ and $\mathbf{p}^r_s \in \mathbb{N}_0$ denote the stoichiometry of each species $s \in \Lambda$), $k^r \in \mathbb{R}_{\geq 0}$ denotes the rate of r and $\mathbf{k}$ denotes the vector of all reaction rates. Given a reaction $r = (\mathbf{r}^r, \mathbf{p}^r, k^r)$, the set of reactants of r is $\{s \in \Lambda \mid \mathbf{r}^r_s > 0\}$ and the set of products of r is $\{s \in \Lambda \mid \mathbf{p}^r_s > 0\}$. In this paper, we focus on the class of *bimolecular* CRNs, where $\sum_{s \in \Lambda} \mathbf{r}^r_s = 2$ and $\sum_{s \in \Lambda} \mathbf{p}^r_s = 2$, for all reactions $r \in \mathcal{R}$.

The dynamical behaviour of bimolecular CRNs can be understood as follows. The set of all possible system states is $X = \mathbb{N}_0^{|\Lambda|}$, where a state $x \in \mathbb{N}_0^{|\Lambda|}$ represents the number of molecules of each species. We denote the number of molecules of species $s \in \Lambda$ at state x by x_s. Given a reaction $r \in \mathcal{R}$ where $\mathbf{r}^r_s = 2$ for some $s \in \Lambda$, the *propensity*[1] of r at x is $k^r_x = k^r \cdot \frac{x_s \cdot (x_s - 1)}{2}$. If, on the other hand, $\mathbf{r}^r_s = \mathbf{r}^r_{s'} = 1$ for some species s, s', the propensity of r is $k^r_x = k^r \cdot x_s \cdot x_{s'}$. The time at which reaction r would fire, once the system enters state $x \in X$, is stochastic and follows an exponential distribution with a rate determined by the reaction's propensity k^r_x. Assuming that reaction r is the first one to fire, the state of the system is updated as $x'_s = x_s - \mathbf{r}^r_s + \mathbf{p}^r_s$ for all $s \in \Lambda$, where x and x' are the current and next states.

An abstraction of CRNs that preserves reachability but does not consider reaction rates or time is given by the *transition system* $T^{\mathcal{C}} = (X, T)$, where the transition relation T is defined as

$$\forall x, x' \in X . \, T(x, x') \leftrightarrow \bigvee_{r \in \mathcal{R}} \bigwedge_{s \in \Lambda} (x_s \geq \mathbf{r}^r_s \wedge x'_s = x_s - \mathbf{r}^r_s + \mathbf{p}^r_s). \qquad (1)$$

In other words, the choice between reactions from $\mathcal{R}$ is non-deterministic but enough molecules of each reactant must be present in state x for the reaction to fire. The transition between states x and x' happens when any reaction $r \in \mathcal{R}$ fires and the number of molecules is updated accordingly. A path $x_0, x_1, \ldots$ of T satisfies $T(x_i, x_{i+1})$ for $i = 0, 1, \ldots$ and, given an initial state x_0 we call state x_f reachable from x_0 if there exists a path $x_0, \ldots, x_f$.

Given a CRN $\mathcal{C}$, let $X_0 \subseteq X$ denote a finite set of initial states and $X_r \subseteq X$ denote the set of states reachable from X_0. Assuming that X_r is finite, $\mathcal{C}$ can be represented as a *continuous time Markov chain* (CTMC) that preserves information about the transition probabilities and rates that determine the stochastic behaviour of the system and the expected execution times. We define a CTMC to be a tuple $\mathcal{M} = (X_r, \pi_0, \mathbf{Q})$, where X_r is a finite set of states, $\pi_0 : X_r \to \mathbb{R}$ is the initial distribution of molecule copy numbers of all species, and $\mathbf{Q} : X_r \times X_r \to \mathbb{R}$ is a matrix of transition propensities. While the set of initial states is not represented explicitly, it is captured through the initial distribution, i.e. $X_0 = \{x \in X_r \mid \pi_0(x) > 0\}$. A CTMC $\mathcal{M}^{\mathcal{C}}$ is constructed from a CRN $\mathcal{C}$ by first determining the set of reachable states, and then evaluating the propensities of each reaction. The $(i, j)^{\text{th}}$ entry of $\mathbf{Q}$, q_{ij}, represents a transition from state x_i to state x_j. Accordingly, q_{ii} is the remaining probability mass,

---

[1] We assume that the reaction volume is 1 to allow for later volume scaling e.g. k^r_x/v is the propensity for a reaction volume equal to v.

equal to $-\sum_{i \neq j} q_{ij}$. The transient probability vector π_t evolves according to $\frac{d\pi_t}{dt} = \pi_t Q$, which is known as the chemical master equation (CME).

Following [13,14], a parametric CTMC (pCTMC) is a CTMC where the reaction rates are parameterised by $\mathbf{k}$, as above. Denote by $\mathcal{P}$ the parameter space, $\mathcal{P} : \mathbb{R}_{\geq 0}^P$, such that $\mathbf{k}$ is instantiated by a parameter point $p \in \mathcal{P}$. Accordingly, given a pCTMC $\mathcal{M}$ and parameter space $\mathcal{P}$, an instantiated pCTMC $\mathcal{M}_p = (X, \pi_0, Q_p)$ is an evaluation at point $p \in \mathcal{P}$.

3 Problem Formulation

The main problem we consider in this paper, which we formalise in this section, is the identification of CRNs that satisfy given properties. Specifically, we are interested in finite reachability properties, which capture a range of interesting CRN behaviours.

Let $\mathcal{C} = (\Lambda, \mathcal{R})$ be a given CRN and $\mathcal{T}^{\mathcal{C}} = (X, T)$ and $\mathcal{M}^{\mathcal{C}} = (X_r, \pi_0, Q)$ denote its transition system abstraction and CTMC representation, as discussed in Sect. 2. Let $\phi : X \to \mathbb{B}$ denote a *state predicate*, constructed using

$$\phi : := E_b$$
$$E_b : := true \mid false \mid E_c \mid \neg E_b \mid E_b \triangleright E_b \text{ where } \triangleright \in \{\wedge, \vee, \Rightarrow, \Leftrightarrow\}$$
$$E_c : := E_a \triangleright E_a \text{ where } \triangleright \in \{<, \leq, =, >, \geq\}$$
$$E_a : := s \in \Lambda \mid c \in \mathbb{Z} \mid E_a \triangleright E_a \text{ where } \triangleright \in \{+, -, *\}.$$

For example, if $\phi := s > 5$, then $\phi(x)$ denotes that $x_s > 5$.

In this paper, we consider *path predicates* $\Phi = (\phi_0, \phi_F)$, which are expressed using two state predicates that must be satisfied at the initial (ϕ_0) and at some final (ϕ_F) state of a path. Let K denote the number of steps we consider.

Definition 1. *Given a finite path $\rho : x_0 \ldots x_K$ of $\mathcal{T}^{\mathcal{C}}$ we say that ρ satisfies path predicate $\Phi = (\phi_0, \phi_F)$, denoted as $\rho \models \Phi$, if and only if $\phi_0(x_0) \wedge \phi_F(x_K)$ evaluates to true and no reactions are enabled in x_K (i.e. x_K is a terminal state).*[2]

We define the probability of Φ, denoted P_Φ, using $\mathcal{M}^{\mathcal{C}}$ as follows. Let $X_0 = \{x \in X \mid \phi_0(x)\}$ denote the set of states that satisfy the initial state predicate. We initialise $\mathcal{M}^{\mathcal{C}}$ with a uniform sample from the states that satisfy ϕ_0, which defines π_0 as

---

[2] We consider terminating computations by enforcing that no reactions are enabled at the state that satisfies ϕ_F. Alternative strategies possible within our approach could consider reaching a fix-point (i.e. the firing of any enabled reaction does not cause a transition to a different state), or reaching a cycle along which ϕ_F is satisfied, to guarantee that the correct output is eventually computed and remains unchanged by any subsequent reactions.

$$\pi_0(x) = \begin{cases} \frac{1}{|X_0|} & \text{if } x \in X_0 \\ 0 & \text{otherwise} \end{cases}$$

Similarly, $X_F = \{x \in X \mid \phi_F(x)\}$ denotes the set of states satisfying the final state predicate.

Definition 2. *The probability of Φ is defined as*

$$P_\Phi = \sum_{x \in X_F} \pi_t(x),$$

where t denotes the maximal time we consider and π_t is the probability vector at time t computed using the CME introduced in Sect. 2. In other words, we define P_Φ as the average probability of the states satisfying ϕ_F at time t.

Note that it is possible to optimise for both speed and accuracy by, for example, defining P_Φ to be the integration of the probability mass of all states satisfying ϕ_F from time 0 to time t.

Problem 1. Given a finite set of path predicates $\{\Phi_0, \ldots, \Phi_n\}$, find a bimolecular CRN $\mathcal{C}$ such that

1. for each Φ_i, there exists a path ρ_i of $\mathcal{T}^\mathcal{C}$, such that $\rho_i \vDash \Phi_i$ and
2. the average probability $\frac{\sum_{i=0}^n P_{\Phi_i}}{n+1}$ defined using $\mathcal{M}^\mathcal{C}$ is maximised.

4   Synthesis and Tuning of CRNs

We solve Problem 1 by addressing each of the two subproblems separately. First, we generate a number of CRNs that satisfy the specifications from Problem 1.1 using a satisfiability modulo theories (SMT)-based approach (Sect. 4.1). The CRNs identified at that point are capable of producing a path that satisfy each path predicate, which addresses Problem 1.1 but they might also include incorrect paths and the probability of correct computations might be low. Therefore, we tune the reaction rates of these CRNs in order to maximise the average probability (discussed in Sect. 4.2), which addresses Problem 1.2

4.1   SMT-Based Synthesis

Here, we present our approach to finding a bimolecular CRN $\mathcal{C}$ that satisfies a specification expressed as path predicates $\{\Phi_0, \ldots, \Phi_n\}$ (Problem 1.1). We address this problem by encoding $\mathcal{T}^\mathcal{C}$ symbolically for any possible bimolecular CRN $\mathcal{C} = (\Lambda, \mathcal{R})$ where $|\mathcal{R}| = M$ and $|\Lambda| = N$ (i.e. the number of species and reactions is given), together with the specification $\{\Phi_0, \ldots, \Phi_n\}$ for some finite number of steps K, as a satisfiability modulo theories (SMT) problem. We then use the SMT solver Z3 [12] to enumerate bimolecular CRNs that satisfy the specification or prove that no such CRNs exists for the given N, M, and K. Finally, we apply an incremental procedure to search for CRNs of increasing complexity (larger N and M) or to provide more complete results by increasing K.

Using Z3's theory of linear integer arithmetic, we represent the stoichiometry of $\mathcal{C}$ as two symbolic matrices $\mathbf{r} \in \mathbb{N}_0^{M \times N}$ and $\mathbf{p} \in \mathbb{N}_0^{M \times N}$ (using integer constraints to prohibit negative integers). Given a reaction $r \in \mathcal{R}$ and species $s \in \Lambda$, $\mathbf{r}_s^r$ ($\mathbf{p}_s^r$) defined in Sect. 2 is now encoded as a symbolic integer. We ensure that only bimolecular CRNs are considered by asserting the constraints $\bigwedge_{i=0}^{M-1} \sum_{j=0}^{N-1} \mathbf{r}_{i,j} = 2$ and $\bigwedge_{i=0}^{M-1} \sum_{j=0}^{N-1} \mathbf{p}_{i,j} = 2$. In addition, we introduce the following constraints.

- We label a subset of the species $\Lambda_I \subseteq \Lambda$ as inputs and assert that $\bigwedge_{s \in \Lambda_I} \bigvee_{r \in \mathcal{R}} \mathbf{r}_s^r > 0$ to ensure all inputs are consumed by at least one reaction.
- We label a subset of the species $\Lambda_O \subseteq \Lambda$ as outputs and assert that $\bigwedge_{s \in \Lambda_O} \bigvee_{r \in \mathcal{R}} \mathbf{p}_s^r > 0$ to ensure all outputs are produced by at least one reaction.
- We assert that $\bigwedge_{r,r' \in \mathcal{R}, r \neq r'} \bigvee_{s \in \Lambda} \mathbf{p}_s^r \neq \mathbf{p}_s^{r'} \vee \mathbf{r}_s^r \neq \mathbf{r}_s^{r'}$ to ensure that two reactions never have the same reactants and products and, therefore, all M reactions are utilised.
- Finally, we assert that $\bigwedge_{r \in \mathcal{R}} \bigvee_{s \in \Lambda} \mathbf{p}_s^r \neq \mathbf{r}_s^r$ to ensure that the firing of each reaction updates the state of the system.

Following an approach inspired by bounded model checking (BMC) [18], we represent the finite path $\rho_i = x_0^i, \dots x_K^i$ for each Φ_i by defining each state as a symbolic vector $x_j^i \in \mathbb{N}_0^N$ and "unrolling" the transition relation of $\mathcal{T}_C$ (i.e. asserting the constraint $T(x_j^i, x_{j+1}^i)$ for each $i = 0 \dots n$ and $j = 0 \dots K-1$). For each path predicate $\Phi_i = (\phi_0, \phi_F)$ and path ρ_i we then assert the constraint $\phi_0(x_0^i) \wedge \phi_F(x_K^i) \wedge \textit{Terminal}(x_K^i)$ according to Definition 1, where $\textit{Terminal}(x) \triangleq \bigwedge_{r \in \mathcal{R}} \bigvee_{s \in \Lambda} x_s < \mathbf{r}_s^r$, i.e. no reactions are possible due to insufficient molecules of at least one reactant.

The parameter K specifies the maximal trajectory length that is considered. The BMC approach is conservative, since computations that require more than K steps (reaction firings) to reach a state satisfying ϕ_F will not be identified. Increasing K leads to a more complete search, and indeed the approach becomes complete for a sufficiently large K determined by the diameter of a system, but also increases the computational burden. To alleviate this, we follow an approach from [11] and consider *stutter* transitions (corresponding to multiple firings of the same reaction in a single step) by using the following modified transition relation definition T_{st} (as opposed to T from Eq. 1)

$$\forall x, x' \in X . \; T_{st}(x, x') \leftrightarrow (\textit{Terminal}(x) \wedge x = x') \vee$$
$$\exists n \in \mathbb{N} . \; \bigvee_{r \in \mathcal{R}} \bigwedge_{s \in \Lambda} (x_s \geq \mathbf{r}_s^r \wedge x_s \geq n \cdot (\mathbf{r}_s^r - \mathbf{p}_s^r) \wedge x_s' = x_s + n \cdot (\mathbf{p}_s^r - \mathbf{r}_s^r)).$$

For any enabled reaction r ($x_s \geq \mathbf{r}_s^r$), T_{st} allows r to fire up to n times in the stutter transition. n is limited by the consumption and production of the species needed for the reaction to fire ($x_s \geq n \cdot (\mathbf{r}_s^r - \mathbf{p}_s^r)$). In many cases, stutter transitions dramatically decreases the required trajectory lengths (K), since multiple copies of the same species can react simultaneously. However, this is not restrictive, since for $n = 1$ the original definition of T is recovered. In addition to such stutter transitions, T_{st} allows self loops at terminal states, and therefore computations that require less than K steps to reach a state satisfying ϕ_F can also be identified.

The encoding strategy described so far allows us to represent CRN synthesis as an SMT-problem and apply an SMT solver such as Z3 [12] to produce a CRN that satisfies the specification or prove that no such CRN exists for the choice of M, N and K. More specifically, a solution CRN $\mathcal{C}$ is represented through the valuation of $\mathbf{r}$ and $\mathbf{p}$, which are extracted from the *model* returned by Z3.

In general, we are interested in enumerating many (or all possible) CRNs for the given class (defined by M, N and K), which ensures that no valid solutions are omitted at that stage. To do so, we apply an incremental SMT-based procedure, where at each step we assert an uniqueness constraint guaranteeing that no previously discovered CRNs are generated. Given a concrete, previously generated CRN $\mathcal{C}' = (\Lambda, \mathcal{R}')$ and the new symbolic CRN $\mathcal{C} = (\Lambda, \mathcal{R})$ we are searching for (both of which are defined using the same species Λ), we define the constraint $DifferentFrom(\mathcal{C}') \triangleq \neg \bigwedge_{r \in \mathcal{R}} \bigvee_{r' \in \mathcal{R}'} r = r'$, where $r = r'$ if and only if $\mathbf{r}_s^r = \mathbf{r}_s^{r'} \wedge \mathbf{p}_s^r = \mathbf{p}_s^{r'}$ for all $s \in \Lambda$. The new CRN $\mathcal{C}$ cannot simply be a permutation of the same reactions[3]. We start by generating a solution $\mathcal{C}'$ (if one exists), asserting the constraint $DifferentFrom(\mathcal{C}')$, and repeating this procedure until the constraints become unsatisfiable, which corresponds to a proof that not additional CRNs exists for the given N, M, and K.

4.2 Tuning CRNs with Parameter Optimisation

Here, we present our approach to optimising the reaction rates for CRNs satisfying $\{\Phi_0, \ldots, \Phi_n\}$. This becomes a parameter synthesis problem over a pCTMC set, analogous to parameter synthesis for a single pCTMC, as studied in [13,14]. In contrast to this work, we aggregate over the multiple input combinations, as specified in Problem 1.2.

To obtain solutions for the probability at a specified time π_t, we used numerical integration of the CME. Specifically, we used the Visual GEC software (http://research.microsoft.com/gec) to encode the CRNs and then integrate the CME for each combination of inputs.

To solve the maximisation problem, we used a Markov chain Monte Carlo (MCMC) method, as implemented in the Filzbach software (http://research.microsoft.com/filzbach). Filzbach uses a variation of the Metropolis-Hastings (MH) algorithm to perform Bayesian parameter inference. The MH algorithm is used to approximate the posterior probability of a parameter set from a hypothesised model taking on certain values, constrained by a likelihood function. The

---

[3] At present, our uniqueness constraint does not consider other CRN isomorphisms but certain species symmetries are broken by the specification Φ_i.

probability of each parameter value is then approximated by constructing a Markov chain of sampled parameter sets, such that a proposed parameter set is accepted with some probability, based on the ratio of the likelihood function evaluated at current and proposal parameter sets. For more information on MCMC methods, see [19]. MCMC methods, such as simulated annealing, have also been shown to efficiently find solutions to combinatorial optimisation problems [20], taking a stochastic search approach similar to the MH algorithm. Stochastic search can provide benefits over gradient-based optimisers by maintaining a nonzero probability of making up-hill moves, protecting against getting stuck in poor local optima. To use Filzbach for providing solutions to optimisation CRN parameters, it is sufficient to encode the argument of Problem 1.2 as a likelihood function. Subsequently, we generate MCMC chains with suitably many burn-in iterations and samples to obtain an approximate optimising parameter set $\mathbf{k}$.

4.3 Calculating Expected Time

To evaluate the temporal performance of a CRN algorithm $\mathcal{C}$, we make use of Markov chain theory to obtain the expected time until a terminal state is reached. This is an exact measure of the expected running time for a given pCTMC with inputs $i \in \mathcal{I}$, as opposed to using the mean of many stochastic simulations [10].

Let $A \subseteq X_r$ be the absorbing states of a pCTMC $\mathcal{M}_p^{\mathcal{C}} = (X, \pi_0, \mathbf{Q}_p)$ and let τ^A be a vector of expected hitting times, corresponding to the expected time of transitioning from a state $x \in X_r$ to A. Then τ^A can be evaluated as the solution to the equations (page 113 of [21])

$$\tau_x^A = 0 \text{ for } x \in A$$
$$-\sum_{x' \in X_r} q_{x,x'} \tau_{x'}^A = 1 \text{ for } x \notin A.$$

Numerical solutions can be obtained by forming a matrix W where the rows and columns of $\mathbf{Q}_p$ corresponding to the terminal states (A) have been removed. Then, τ^A is the solution to $W\tau^A = \mathbf{1}$, where $\mathbf{1}$ is the vector of 1's. Numerical solutions can be obtained using Gaussian elimination.

Note that the time complexity analysis of CRNs typically assumes a volume n equal to the maximum number of molecules in the system at any time [8] (equivalent to parallel time in PPs [10]). This volume can be included by dividing each propensity by n before calculating expected time (see Sect. 2). In the case of bimolecular CRNs this is equivalent to multiplying τ^A by n.

5 Case Studies

5.1 Approximate Majority

Approximate Majority is one of the most analysed functions in distributed computing. It is the approximate version of the majority problem, which cannot be exactly computed by bimolecular CRNs (or population protocols) with less than 4 species [22]. For CRNs with 2 and 3 species there are known optimal (in terms of reaction firings) approximate algorithms [15,16].

We specify the majority problem using the path predicate (see Sect. 2): $\Phi_{AM}(a,b) := (\phi_0(a,b), \phi_F(a,b))$, where

$$\phi_0(a,b) := \begin{cases} A = a \wedge B = b & \text{if } N = 2, \\ A = a \wedge B = b \wedge X = 0 & \text{if } N = 3 \end{cases}$$

$$\phi_F(a,b) := \begin{cases} A^m_{a,b} & \text{if } a > b \\ B^m_{a,b} & \text{if } a < b \\ A^m_{a,b} \vee B^m_{a,b} & \text{otherwise} \end{cases}$$

$$\text{where } A^m_{a,b} := A = a + b \wedge B = 0 \text{ and}$$
$$B^m_{a,b} := A = 0 \wedge B = a + b$$

We used inputs $a, b \in [1 \ldots 5]^2 \cup [6 \ldots 10]^2$ for both optimisation and synthesis. We applied the SMT approach to identify all CRNs with 2 to 4 reactions and 2 or 3 species that satisfy Φ_{AM} for $K \leq 5$ stutter steps (for N species and M reactions, there are $\binom{N^2(N^2-1)}{M}$ total possible CRNs). We used a short optimisation (20 burn-in, 20 samples) and sorted these solutions by the value of $P_{\Phi_{AM}}$ for each. We then applied a longer optimisation (700 burn-in, 700 samples) to the top 10 CRNs (Fig. 1).

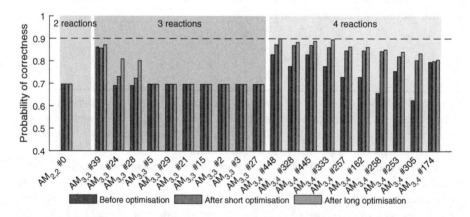

Fig. 1. Performance of approximate majority circuits. The SMT-based method was applied to the approximate majority specification for CRNs with 2, 3 and 4 reactions. For each category, the top 10 CRNs satisfying Φ_{AM} are ordered by their average probability after a short optimisation (20 burn-in, 20 samples; red bars). A longer optimisation (700 burn-in, 700 samples; green bars) was also performed. We also show the average probabilities before optimisation (all rates equal to 1.0; blue bars). The dashed line is the average probability of CRN $AM_{3,4}$ #448 after the longer optimisation, 0.8999, the maximum average probability in this trail.

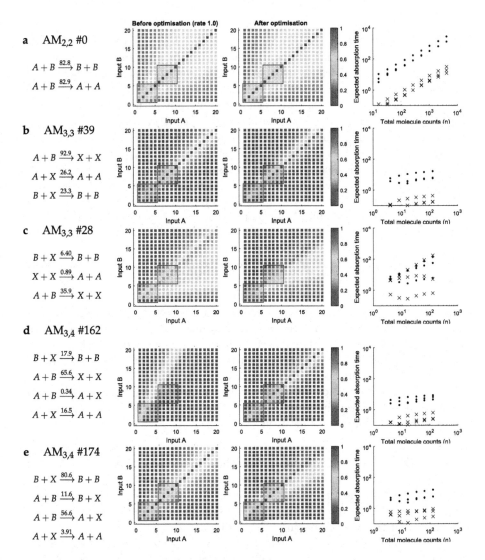

Fig. 2. Response of Approximate Majority algorithms to varied inputs. For each input combination, specified as initial copies of species A and species B, the probability that both have the correct molecule count after 100 time units is reported. Results are shown for a variety of networks that performed well following optimisation (see Fig. 1). The performance of each CRN is compared both before optimisation (all rates equal to 1.0; left panels) and after long optimisation (central panels). The grey boxes show the input ranges used for both generation and optimisation. The expected time until the CTMC reaches a terminal state is calculated for varying total molecule counts (n) (right panels). These times consider rates scaled as if occurring in a volume n (see Sect. 4.3). The completion times for three alternative initial configurations (initial copies of A were 10%, 60% and 90% of n respectively) were calculated, illustrating minor differences in circuit completion times ($\times$ marks systems using optimised rates and $\bullet$ marks systems using 1.0 for all rates).

Using our approach, we found 1 CRN with 2 reactions and 2 species, the known *direct competition* (DC) network [23] (Fig. 2a). Out of 59,640 possible CRNs with 3 species and 3 reactions, the SMT solver found 39 CRNs where Φ_{AM} was satisfied, 2 of which with probability over 0.696 after the short optimisation (see Fig. 1). These two networks ($AM_{3,3}$ #24 and $AM_{3,3}$ #28) are the dual of each other and behave asymmetrically but perform well owing to a compensatory asymmetric parameterisation (Fig. 2c). One might expect that we should discover the known approximate majority circuit [15,17], (see Fig. 2b). However, this CRN does not satisfy the specification Φ_{AM} since, for input ($A = 1, B = 1, X = 0$) the network terminates in the state ($A = 0, B = 0, X = 2$) and thus fails to make a decision. If we remove this single problematic input from the specification Φ_{AM}, then this CRN is indeed discovered. We include it for comparison as $AM_{3,3}$ #39. Note that it scores a 0 on inputs $A = 1$, $B = 1$.

By increasing the number of reactions to 4, the SMT solver found 515 satisfying networks out of the 1,028,790 possible ones. The top 5 networks, $AM_{3,4}$ #448, #328, #445, #333, and #257 have the same rules as the 3 reaction network $AM_{3,3}$ #39 but each has a different 4th reaction. The network $AM_{3,4}$ #162 had a lower performance than $AM_{3,3}$ #39 before optimisation and was almost as good following optimisation. This network was also asymmetric, with a corresponding asymmetric parameterisation after optimisation (Fig. 2d). The known 4 reaction network $AM_{3,4}$ #174 [17] (Fig. 2e) is also identified in 10th position.

Finally, we analysed the expected time until termination for each circuit, using the procedure in Sect. 4.3 (right-hand panels of Fig. 2). Note that Definition 2 does not reward circuits that reach a high probability before the final time $t_f = 100$. However, in nearly all cases, the estimated hitting time of each system was improved by optimisation.

Computation Times. The computation times of our procedure depend on the size of the circuit (M and N), length of considered computations (K) and exact specification Φ (including the number of given path predicates). We illustrate the computation times required for the SMT-based synthesis part of our approach with the majority decision-making CRNs (Fig. 3).

To determine how the CME calculation used in our method scales with molecular copy numbers, we first ran calculations of the CME for the established 3-reaction approximate majority CRN (system $AM_{3,3}$ #39). The calculation was initialised with $0.6n$ copies of A and $0.4n$ copies of B, and all rates were set to 1. As increasing the copy number decreases the simulation time interval over which there are transient dynamics, we integrated the CME over the time interval $\left[0, \frac{100}{n}\right]$, where n is the total copy number. We calculated transient probabilities at 500 output points, with $n \in [10, 1000]$. This led to state-spaces of varying size, up to 10^6, with all calculations completing within 7200 s (2 h) (Fig. 4). Smaller examples took only a few seconds.

We can approximate the total run-time for parameter tuning as a function of the number of iterations of the MCMC algorithm and the number of input combinations assessed. For example, doing 200 iterations over 10 input

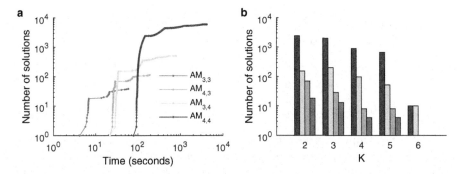

Fig. 3. Computation times for the SMT-based synthesis of majority decision-making CRNs. Panel (a) shows the time required to generate a number of solutions (candidate CRNs) for Φ_{AM} for N species and M reactions (denoted $AM_{N,M}$) for $N, M \in \{3,4\}^2$. The computation was halted after 2 h. Panel (b) shows the number of solutions found as K (the length of considered computations with stutter transitions) increases.

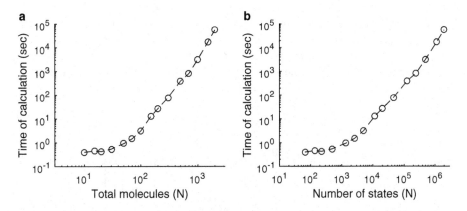

Fig. 4. Transient probability calculation times for CRN $AM_{3,3}$ #39. Times indicated include the enumeration of the state-space, construction of a sparse matrix, then numerical integration in the interval $[0, \frac{100}{n}]$, where n is the total molecule count. A single calculation was conducted for each value of n.

combinations which all have below 30 total molecules ($\lesssim 1$ s each) suggests a tuning procedure of no more than 2,000 s.

5.2 Division

Division is a non-semi-linear function and therefore it cannot be stably computed by CRNs [8]. However, CRNs have been proposed that might implement the calculation of a ratio [24], which allows plants to ration starch reserves during seasonally changing nights.

We specify the division problem using the path predicate (see Sect. 2):

$$\Phi_{\text{Div}}(a, b) := (\phi_0(a, b), \phi_F(a, b), \text{ where}$$

$$\phi_0(a, b) := \begin{cases} A = a \wedge B = b \wedge X = 0 & \text{if } N = 3 \\ A = a \wedge B = b \wedge X = 0 \wedge Y = 0 & \text{if } N = 4 \end{cases}$$

$$\phi_F(a, b) := X = \left\lfloor \frac{a}{b} \right\rfloor$$

We chose the input ranges $a, b \in [1, \ldots, 10]^2$ for synthesis and optimisation to give diverse selection of responses and to reinforce that $\lfloor \frac{a}{b} \rfloor = 0$ when $a < b$. We applied the SMT approach to CRNs that satisfied Φ_{Div} with $K < 20$ (without stutter transitions). For 3 species and 3 reactions, 22 CRNs were discovered. For 4 species and 3 reactions, 34 CRNs were discovered. For 4 species and 4 reactions the first 105 CRNs were discovered. Of these, only one CRN $DIV_{4,3}$ # 29 exceeded an average probability of 0.5, though in most cases, optimisation improved performance substantially (Fig. 5). For many of the generated circuits, high performance was observed only for $b > a$, which should always evaluate to 0, with poor performance for the nonzero output cases of $a > b$ (Fig. 6a, b). Note that $Div_{4,3}$ #29 is so far the top scoring divider CRN in this class. Clearly, none of these circuits can be considered as *good* algorithms for computing division, though our procedure was able to detect some very simple yet mediocre circuits in an automated way. It is possible that better circuits will be found by considering CRNs with more reactions, species, and longer computation paths.

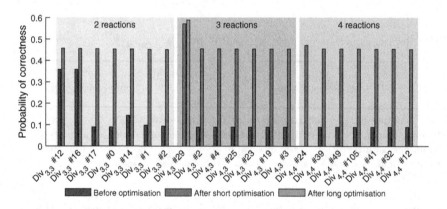

Fig. 5. Performance of division circuits. The SMT-based method was applied to the division specification for CRNs with N species and M reactions for $N, M \in \{(3,3), (3,4), (4,4)\}$. This figure shows the optimisation results for the top 7 CRNs in each category. The results are ranked and sorted by their average probability of being correct in the grey shaded zone after being optimised for 50 MCMC sample and burn-in steps (red bars). If a CRN scored an average probability of over 0.5 then it was optimised for a further 200 MCMC burn-in and sample steps. The average probability is shown for satisfying CRNs before optimisation (all rates equal to 1.0; blue bars).

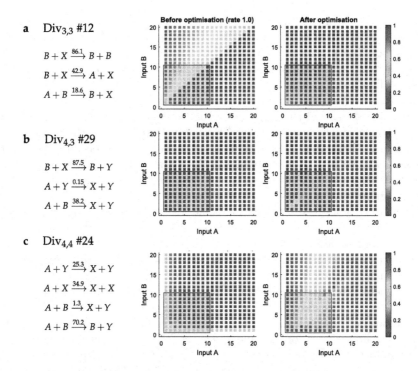

a $\text{Div}_{3,3}$ #12

$$B + X \xrightarrow{86.1} B + B$$
$$B + X \xrightarrow{42.9} A + X$$
$$A + B \xrightarrow{18.6} B + X$$

b $\text{Div}_{4,3}$ #29

$$B + X \xrightarrow{87.5} B + Y$$
$$A + Y \xrightarrow{0.15} X + Y$$
$$A + B \xrightarrow{38.2} X + Y$$

c $\text{Div}_{4,4}$ #24

$$A + Y \xrightarrow{25.3} X + Y$$
$$A + X \xrightarrow{34.9} X + X$$
$$A + B \xrightarrow{1.3} X + Y$$
$$A + B \xrightarrow{70.2} B + Y$$

Fig. 6. Response of Division algorithms to varied inputs. For each input combination, specified as initial copies of species A and species B, the probability that the molecule count of X is $\lfloor A/B \rfloor$ after 100 time units is reported. Results are shown for the top network in each combination of species and reactions (see Fig. 5). The performance of each CRN is compared both before optimisation (all rates equal to 1.0; left panels) and after optimisation (right panels).

6 Discussion

In this paper, we presented a computational approach for the synthesis and parameter tuning of CRNs, given a specification of the system's correctness. We focused on the sub-class of bimolecular CRNs due to their importance as representations of various molecular algorithms and population protocols. However, our approach is more general and could also be applied directly to the synthesis of CRNs from other classes (e.g. unimolecular, trimolecular, etc.), which are defined through different stoichiometry constraints. The CRNs we synthesize can be converted into equivalent physical implementations, for example using DNA strand displacement (DSD) [2,3]. However, our approach could also be applied directly to synthesize DSD systems through additional structural constraints. This could lead to simpler designs than the ones obtained through direct translation of CRNs.

We considered simple reachability properties defined in terms of predicates on the initial and final states of a computation which are sufficient to express various logical and arithmetic functions and operations. More general specifications, for example where intermediate states along computations are specified, are also currently possible within our approach but extensions to more expressive languages, such as the probabilistic temporal logics used with other methods [14], remains a direction for future work.

An alternative approach to the problem of realising arbitrary behaviour in biochemical systems is to use directed evolution [25,26] *In silico* evolutionary search strategies might scale to larger CRNs and address the synthesis and parameter optimisation sub-problems using a single, combined procedure. However, this comes at the cost of completeness, where the absence of a solution does not mean a solution does not exist. In contrast, our method addresses the sub-problems separately and uses the SMT solver and theorem prover Z3 to identify CRNs that satisfy a given specification (kinetics are ignored at this first stage). Since the results provided by Z3 are complete (for a sufficiently large K), the termination of the procedure with no solutions is a "proof" that no CRNs exist in the given class. Thus, besides providing a practical tool for the identification of CRNs with given behaviour, the completeness property means our approach could also help explore the theoretical limits of CRN computation (e.g. no CRNs with less than M species and N reactions that compute a given function exists). For many applications, elements of our method could be complementary with evolutionary algorithms. For example, the exact CTMC methods we use to assess the probability of correct computations in a given CRN could provide a useful fitness function for evolutionary search, compared to alternative approximate methods based on stochastic simulation.

The fully automated generation of "good" CRNs is a challenging problem and certain scalability limitations of our current method must be addressed to provide a more complete solution. Firstly, the SMT-based synthesis procedure we propose may represent large or infinite state spaces and handle systems with large molecule numbers. However, currently this method is limited to relatively small CRNs with few reactions, species, and which have short computation paths. Secondly, the CTMC methods we apply require an explicit representation of the state space, which must be finite (which is always the case for biomolecular CRNs initialised with a finite number of molecules) and contain few reachable states — this makes the method suitable for systems involving relatively few species and numbers of molecules. To circumvent the need for an explicit representation of the state space, stochastic dynamical behaviour could be approximated by averaging multiple trajectories from Gillespie's stochastic simulation algorithm [27], using fluid or central limit approximations [28], or using ordinary differential equations. Depending on the specification, and the nature of the CRN, some of these approaches might be appropriate, but none are free of their own documented limitations. Finally, the large number of solutions identified at the synthesis stage of our approach makes the parameter tuning phase challenging and indicates that additional constraints describing more accurately the structure and dynamics of "good" solutions could improve the method.

For tuning reaction rates, alternative cost functions could be used that reward solutions that are "nearly" correct, e.g. using a mean-squared error. This would be most appropriate in high copy number situations, where a precise number of molecules is not integral. Our approach is more appropriate for systems operating at low copy numbers, offering an exact characterisation of the probability that a specific predicate is satisfied. Our results were shown for calculations at $t_f = 100$ time units, a transient probability, rather than at the stationary distribution. While the selection of t_f is subjective, it allows a circuit programmer to specify how long they are willing to wait for a computation. Circuits that reach high probability at $t > t_f$ will not be rewarded. However, a natural extension to the presented method would be to reward circuits that reach high probability at $t < t_f$, both imposing an upper bound on time and optimising within that range. This could be achieved by integrating our metric over the interval $[0, t_f]$.

Automating the search for CRNs that compute the solution for a specified problem would be beneficial to both theoretical and experimental molecular programmers. Our method can be used to show the existence or absence of CRNs of a certain size and also suggest CRNs that can be tuned for a specific input range, and so become candidate designs for experimental construction. Prior to construction, more in-depth analysis of the candidate CRNs produced is beneficial, including parameter sensitivity/robustness analysis and bifurcation analysis (where appropriate). Future work could also incorporate notions of robustness into the proposed method, for example by using interval-based methods [14]. Our results illustrate the potential of this approach on several examples, including the majority and division functions discussed here.

Acknowledgements. We thank Dan Alistarh and Luca Cardelli for helpful discussions on the development and applications of our methodology.

References

1. Wilhelm, T.: The smallest chemical reaction system with bistability. BMC Syst. Biol. **3**, 90 (2009)
2. Soloveichik, D., Seelig, G., Winfree, E.: DNA as a universal substrate for chemical kinetics. PNAS **107**, 5393–5398 (2010)
3. Chen, Y.-J., Dalchau, N., Srinivas, N., Phillips, A., Cardelli, L., Soloveichik, D., Seelig, G.: Programmable chemical controllers made from DNA. Nat. Nanotechnol. **8**(10), 755–762 (2013)
4. Fujii, T., Rondelez, Y.: Predator-prey molecular ecosystems. ACS Nano. **7**(1), 27–34 (2013)
5. Kim, J., Winfree, E.: Synthetic in vitro transcriptional oscillators. Mol. Syst. Biol. **7**(1), 465 (2011)
6. Cook, M., Soloveichik, D., Winfree, E., Bruck, J.: Programmability of chemical reaction networks. In: Condon, A., Harel, D., Kok, J.N., Salomaa, A., Winfree, E. (eds.) Algorithmic Bioprocesses. Natural Computing Series, pp. 543–584. Springer, Heidelberg (2009)

7. Angluin, D., Aspnes, J., Diamadi, Z., Fischer, M.J., Peralta, R.: Computation in networks of passively mobile finite-state sensors. Distrib. Comput. **18**(4), 235–253 (2006)

8. Chen, H.-L., Doty, D., Soloveichik, D.: Deterministic function computation with chemical reaction networks. In: Stefanovic, D., Turberfield, A. (eds.) DNA 2012. LNCS, vol. 7433, pp. 25–42. Springer, Heidelberg (2012)

9. Angluin, D., Aspnes, J., Eisenstat, D.: Stably computable predicates are semilinear. PODC **2006**, 292–299 (2006)

10. Angluin, D., Aspnes, J., Eisenstat, D.: Fast computation by population protocols with a leader. In: Dolev, S. (ed.) DISC 2006. LNCS, vol. 4167, pp. 61–75. Springer, Heidelberg (2006)

11. Yordanov, B., Wintersteiger, C.M., Hamadi, Y., Phillips, A., Kugler, H.: Functional analysis of large-scale DNA strand displacement circuits. In: Soloveichik, D., Yurke, B. (eds.) DNA 2013. LNCS, vol. 8141, pp. 189–203. Springer, Heidelberg (2013)

12. de Moura, L., Bjørner, N.S.: Z3: an efficient SMT solver. In: Ramakrishnan, C.R., Rehof, J. (eds.) TACAS 2008. LNCS, vol. 4963, pp. 337–340. Springer, Heidelberg (2008)

13. Han, T., Katoen, J., Mereacre, A.: Approximate parameter synthesis for probabilistic time-bounded reachability. In: Real-Time Systems Symposium, pp. 173–182, IEEE (2008)

14. Češka, M., Dannenberg, F., Kwiatkowska, M., Paoletti, N.: Precise parameter synthesis for stochastic biochemical systems. In: Mendes, P., Dada, J.O., Smallbone, K. (eds.) CMSB 2014. LNCS, vol. 8859, pp. 86–98. Springer, Heidelberg (2014)

15. Angluin, D., Aspnes, J., Eisenstat, D.: A simple population protocol for fast robust approximate majority. Distrib. Comput. **21**(2), 87–102 (2008)

16. Perron, E., Vasudevan, D., Vojnovic, M.: Using three states for binary consensus on complete graphs. In: IEEE Infocom 2009, IEEE Communications Society (2009)

17. Cardelli, L.: Morphisms of reaction networks that couple structure to function. BMC Syst. Biol. **8**(1), 84 (2014)

18. Biere, A., Cimatti, A., Clarke, E., Zhu, Y.: Symbolic model checking without BDDs. In: Cleaveland, W.R. (ed.) TACAS 1999. LNCS, vol. 1579, pp. 193–207. Springer, Heidelberg (1999)

19. Robert, C.P., Casella, G.: Monte Carlo Statistical Methods, 2nd edn. Springer, New York (2004)

20. Kirkpatrick, S., Gelatt, C.D., Vecchi, M.P.: Optimization by simulated annealing. Science **220**(4598), 671–680 (1983)

21. Norris, J.R.: Continuous-time Markov Chains. Cambridge University Press, Cambridge (1997)

22. Mertzios, G.B., Nikoletseas, S.E., Raptopoulos, C.L., Spirakis, P.G.: Determining majority in networks with local interactions and very small local memory. In: Esparza, J., Fraigniaud, P., Husfeldt, T., Koutsoupias, E. (eds.) ICALP 2014. LNCS, vol. 8572, pp. 871–882. Springer, Heidelberg (2014)

23. Cardelli, L., Csikász-Nagy, A.: The cell cycle switch computes approximate majority. Sci. Rep. **2**(656) (2012)

24. Scialdone, A., Mugford, S.T., Feike, D., Skeffington, A., Borrill, P., et al.: Arabidopsis plants perform arithmetic division to prevent starvation at night. eLife **2**, e00669 (2013)

25. Soyer, O.S., Bonhoeffer, S.: Evolution of complexity in signaling pathways. PNAS **103**(44), 16337–16342 (2006)

26. Dinh, H., Aubert, N., Noman, N., Fujii, T., Rondelez, Y., Iba, H.: An effective method for evolving reaction networks in synthetic biochemical systems. IEEE Trans. Evol. Comput. **19**(3), 374–386 (2014)
27. Gillespie, D.: Exact stochastic simulation of coupled chemical reactions. J. Phys. Chem. **81**, 2340–2361 (1977)
28. Ethier, S.N., Kurtz, T.G.: Markov Processes: Characterization and Convergence, vol. 282. Wiley, New York (2009)

Universal Computation and Optimal Construction in the Chemical Reaction Network-Controlled Tile Assembly Model

Nicholas Schiefer and Erik Winfree[✉]

California Institute of Technology, Pasadena, CA 91125, USA
winfree@caltech.edu

Abstract. Tile-based self-assembly and chemical reaction networks provide two well-studied models of scalable DNA-based computation. Although tile self-assembly provides a powerful framework for describing Turing-universal self-assembling systems, assembly logic in tile self-assembly is localized, so that only the nearby environment can affect the process of self-assembly. We introduce a new model of tile-based self-assembly in which a well-mixed chemical reaction network interacts with self-assembling tiles to exert non-local control on the self-assembly process. Through simulation of multi-stack machines, we demonstrate that this new model is efficiently Turing-universal, even when restricted to unbounded space in only one spatial dimension. Using a natural notion of program complexity, we also show that this new model can produce many complex shapes with programs of lower complexity. Most notably, we show that arbitrary connected shapes can be produced by a program with complexity bounded by the Kolmogorov complexity of the shape, without the large scale factor that is required for the analogous result in the abstract tile assembly model. These results suggest that controlled self-assembly provides additional algorithmic power over tile-only self-assembly, and that non-local control enhances our ability to perform computation and algorithmically self-assemble structures from small input programs.

1 Introduction

Biological systems are capable of remarkable self-organization directed by information-carrying molecules and the complex biochemical networks that interpret them. Even more remarkably, these systems are able to modify themselves and reconfigure their structure in response to changes in the surrounding environment. In many areas of nanotechnology, we seek to emulate biological systems by implementing self-assembly processes and dynamical systems at the nanoscale.

Because of its relatively rigid, well-understood structure and the specificity of Watson-Crick hybridization, DNA is a common substrate for work in nanotechnology. So far, work in the field has been loosely divided into two classes: structural DNA nanotechnology, which involves self-assembly of small DNA subunits

© Springer International Publishing Switzerland 2015
A. Phillips and P. Yin (Eds.): DNA 2015, LNCS 9211, pp. 34–54, 2015.
DOI: 10.1007/978-3-319-21999-8_3

into larger structures [18,25], and dynamic DNA nanotechnology, which seeks to implement the behavior of dynamical systems through fluctuating quantities of chemical species [34]. Both classes of DNA nanotechnology have been explored extensively, now offering scalable methods for engineering complex nanoscale structures from small components [3,14,21,23,31] and both analog and digital circuits for a variety of tasks [7,20,24,32,35]. Furthermore, both classes have well-studied theoretical models, including a variety of self-assembly models based on Wang tiling [11,16] and models of abstract chemical reaction networks for chemical dynamics [6,8,27,28].

Despite the well-established results in these fields, little theoretical work has considered interactions between structural and dynamic DNA nanotechnology, suggesting that current theoretical models do not capture the full computational power of biomolecular systems. In biological systems, structure influences dynamics, and dynamics influences structure: the two are inextricably linked together. Recently, work by Zhang et al. [33] proposed and experimentally demonstrated a method for controlling the formation of DNA nanotubes from double-crossover tiles using an upstream catalytic circuit implemented as a DNA strand displacement system. However, there is currently little theoretical understanding of the computational power of interacting dynamic and structural biomolecular computing systems.

Some hints come from studies of the computational power of chemical systems involving linear polymers that can store information, which in theory can perform efficient and error-free Turing-universal computation [4,5,19]. In particular, a plausible theoretical implementation of Turing-universal stack machines using dynamic DNA nanotechnology showed that, at least for linear polymers, DNA strand displacement systems can control the assembly and disassembly of nanostructures in a very general and programmable way [19].

Here, we are interested in the ability of biomolecular systems to implement computation and construction tasks in two dimensions (or more): for the former, our goal is to perform a computation and report the answer, while in the latter, our goal is produce a particular nanostructure. Although tile self-assembly permits Turing-universal computation [22] and Kolmogorov-optimal construction (up to scale) [29], the assembly logic in tile self-assembly is local; only the immediate surroundings of a tile can influence its binding. In contrast, chemical reaction networks are usually formulated with a "well-mixed" assumption under which chemical species have no position within the reaction vessel. Although this allows highly non-local information transfer, it precludes the possibility of assembling large complexes. Consequently, we aim to leverage the non-local information transfer offered by chemical reaction networks to exercise non-local control over a two-dimensional (or, in principle, three-dimensional) tile assembly process.

Thus, we introduce the chemical reaction network-controlled tile assembly model (CRN-TAM), a formal model of molecular computing that uses chemical reaction networks to provide non-local control over a tile self-assembly process. In doing so, we formalize and generalize the type of biomolecular computing systems demonstrated experimentally by Zhang et al. [33] and explored theoretically by Qian et al. [19], allowing us to reason mathematically about the capabilities of such systems in comparison to other models of molecular programming. We show that the CRN-TAM subsumes models of stochastic chemical reaction networks and the abstract tile assembly model, and we establish a number of useful "building blocks" for CRN-TAM programs.

Through the Turing-universality of the aTAM, we demonstrate that the CRN-TAM is Turing-universal. Furthermore, we show that the CRN-TAM permits the efficient construction of multi-stack machines, proving that the CRN-TAM is also Turing-universal when restricted to unbounded space in only one spatial dimension, unlike other models of tile-based self-assembly.

Using a natural notion of program complexity, we then turn to bounding the complexity of a minimal CRN-TAM program that constructs a specified algorithmic shape. By explicit construction, we show that there is a CRN-TAM program that constructs every shape $\mathcal{S}$ at scale 2, with complexity bounded by the Kolmogorov complexity of $\mathcal{S}$. We show that this bound is tight by providing a matching lower bound.

2   Defining the CRN-TAM

We begin by outlining a formal definition of the chemical reaction network-controlled tile assembly model and providing a number of useful definitions.

Definition 1. *A* tile *is an oriented square with a bond on each side; the bond positions are called "north," "south," "east," and "west." Each bond has a distinct label and a strength, which is a non-negative integer. Formally, a bond is a tuple (ℓ, s) with label ℓ and strength $s \in \mathbb{N}$. For compactness, we often express a bond $(\ell, 1)$ evocatively as $-_\ell$ and a bond $(\ell, 2)$ as $=_\ell$. A tile is a four-tuple $\boxed{t} = (N, E, S, W)$ of bonds for the north, east, south, and west sides, respectively. Throughout this paper, tiles are denoted by symbols surrounded by boxes, as above.*

Definition 2. *An* assembly *is a function $A : \mathbb{Z}^2 \to (T \cup \{\varepsilon\})$ that gives the type of tile that occupies each site of the 2D lattice, where ε corresponds to an empty site. If $A(x, y) = \varepsilon$, then the site is said to be unoccupied, since there is no tile there. To be a valid assembly at temperature τ, A must satisfy these properties:*

- *The origin must be occupied by a tile $A(0, 0) \neq \varepsilon$, which we call the seed of the assembly.*
- *The occupied sites of the assembly must be connected.*
- *The total binding strength of each tile in the assembly is at least τ.*

Throughout this paper, assemblies are denoted by symbols surrounded by double boxes, e.g. $\boxed{\boxed{A}}$, or shown in a different color from tiles.

The definitions given in Definitions 1 and 2 are identical to those in previous models of tile-based self-assembly derived from the abstract tile assembly model [2,16,22,29]. Although it was implicit in previous "single-crystal" models such as the aTAM, here we explicitly distinguish between free tiles in solution and growing assemblies, even those that contain only a single tile. This ensures formally that free tiles only attach to "activated" assemblies, and not to each other, which is convenient for avoiding issues of spontaneous nucleation and essential for uniform treatment of the "removal signal" reactions described below. Further, it provides a natural way for our model to allow for multiple crystals growing within the same system.

In tile self-assembly, a molecular program is specified by a set T of tiles and their associated bond strengths, an initial seed tile, and a temperature. Analogously, we can define the structural form of a CRN-TAM program:

Definition 3. *A program under the chemical reaction network-controlled tile assembly model is a tuple (S, T, R, τ, I) where*

- *S is a finite set of identified signal species.*
- *T is a finite set of tuples $\left(\boxed{t}, t^*\right)$, where $\boxed{t}$ is a tile and t^* is either ε or some signal species in S. The species t^*, if it exists, is called the removal signal for tile $\boxed{t}$. No tile may appear in more than one tuple.*
- *R is a set of reactions, each of the form:*
  - *$A + B \xrightarrow{k} C + D$ for signals $A, B, C, D \in \{\varepsilon\} \cup S$. These are the "normal" CRN reactions.*
  - *$A + \boxed{T} \xrightarrow{k} C + D$ for signals $A, C, D \in \{\varepsilon\} \cup S$ and tile $\boxed{T}$. These are tile deletion reactions.*
  - *$A + B \xrightarrow{k} \boxed{T} + C$ or $A + B \xrightarrow{k} \boxed{T} + \boxed{T'}$ for signals $A, B, C \in \{\varepsilon\} \cup S$ and tiles $\boxed{T}$ and $\boxed{T'}$. These are tile creation reactions.*
  - *$A + \boxed{T} \xrightarrow{k} B + \boxed{T'}$ for signals $A, B \in \{\varepsilon\} \cup S$ and tiles $\boxed{T}$ and $\boxed{T'}$. These are tile relabelling reactions.*
  - *$A + \boxed{x} \xrightarrow{k} \boxed{\boxed{x}} + X^*$, where $A \in \{\varepsilon\} \cup S$ and $\left(\boxed{x}, X^*\right) \in T$. This tile activation reaction converts a free tile into the seed of a new assembly.*
  - *$\boxed{\boxed{x}} + X^* \xrightarrow{k} A + \boxed{x}$, where $A \in \{\varepsilon\} \cup S$ and $\left(\boxed{x}, X^*\right) \in T$. This tile deactivation reaction converts a seed tile assembly into a free tile.*

  In all of these reactions, k is some rate constant. All of the constructions in this paper are independent of rate constant, and so it is often omitted for notational simplicity. In all cases where a rate constant is omitted, it can be assumed to be 1. When any reactant or product is taken to be ε, the interpretation is that the reactant or product does not exist; for example, a reaction $A + \varepsilon \xrightarrow{k} \varepsilon + D$ is just $A \xrightarrow{k} D$.
- *$\tau \in \mathbb{N}$ is the temperature, or minimum binding strength, typically 0, 1, or 2.*
- *I is an initial state, which is a multiset of tiles and signals that are initially present. Often we will treat I as a function $I : (S \cup T) \to \mathbb{N}$ where $I(z)$ is the count of species z in the multiset. No assemblies are initially present.*

The elements of the set S of signal species are analogous to "normal" species in a chemical reaction network and the set T of tiles is analogous to an aTAM tile

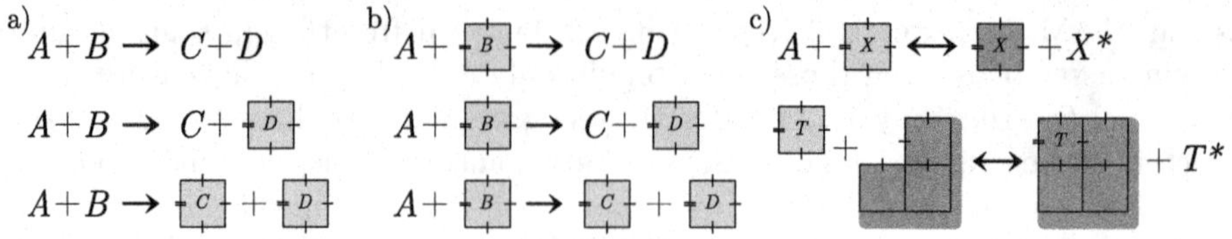

a)
$$A+B \rightarrow C+D$$

$$A+B \rightarrow C+ \boxed{D}$$

$$A+B \rightarrow \boxed{C} + \boxed{D}$$

b)
$$A+ \boxed{B} \rightarrow C+D$$

$$A+ \boxed{B} \rightarrow C+ \boxed{D}$$

$$A+ \boxed{B} \rightarrow \boxed{C} + \boxed{D}$$

c)
$$A+ \boxed{X} \leftrightarrow \boxed{X} +X^*$$

Fig. 1. Example reactions for a CRN-TAM program. (a) Normal chemical reaction network reactions and tile creation reactions. (b) Tile deletion and relabelling reactions. (c) Tile activation, deactivation, addition, and removal reactions.

set (except in that tile concentrations are held constant in the aTAM, while in the CRN-TAM discrete counts of tiles are tracked and may change as reactions proceed). As in the aTAM, tiles may interact with assemblies to form larger structures. However, in the CRN-TAM, each assembly step is accompanied by the release of the tile's associated removal species, and the reaction may be reversible if the removal species is not ε. As a result, the behavior of a CRN-TAM program will be dictated not only by the explicitly specified reactions R as above, but also the *tile addition and removal reactions*:

Definition 4. *A tile addition reaction has the form*

$$\boxed{\boxed{\alpha}} + \boxed{t} \xrightarrow{1} \boxed{\boxed{\beta}} + t^*$$

wherever α and β are valid assemblies that differ by exactly one tile, $\boxed{t}$, that is in β but not in α, where the tuple $\left(\boxed{t}, t^* \right) \in T$. Since β is valid, $\boxed{t}$ formed new bonds with total strength at least τ. The corresponding removal reaction

$$\boxed{\boxed{\beta}} + t^* \xrightarrow{1} \boxed{\boxed{\alpha}} + \boxed{t}$$

may occur only when $\boxed{t}$ is bound by exactly *strength τ.*

We add the condition that a tile removal can only occur if the tile is bound with strength exactly τ based on the principle that reversible reactions should be roughly energetically balanced. As a side effect, it prevents a removal signal from "ripping out" a tile from the middle of an assembly, and thus enforces that only tiles at the boundary can be removed, which corresponds naturally with tile removal in the kinetic tile assembly model.

Together, the contents of a reaction vessel—including free species and assemblies—completely specify the state of a CRN-TAM program at any point: A non-exhaustive sample of allowed reaction types is illustrated in Fig. 1.

Definition 5. *A state L of a CRN-TAM program P is a multiset of signals, tiles, and assemblies.*

Definition 6. *The* propensity *of a reaction is the product of its rate constant and the count of each of its reactants. The* possible reactions *of a state L of a CRN-TAM program $P = (S, T, R, \tau, I)$ are all of the reactions in R with non-zero propensity and all tile addition or removal reactions with non-zero propensity.*

Over time, the program evolves from the initial state according to stochastic Gillespie dynamics: that is, reactions occur at a rate proportional to their current propensity [13]. The time evolution of the state of a CRN-TAM program therefore forms a continuous-time Markov chain.

Definition 7. *An assembly is* terminal *with respect to a state if there can be no possible tile addition or removal reactions involving that assembly in the future.*

While proving that an assembly is terminal can occasionally be done by examination of just the assembly itself (showing that there is no location where a tile may be added or removed, whether or not the tile or removal signal exists in solution), showing that an assembly is terminal is generally undecidable.

Definition 8. *Although the time evolution of the state of a CRN-TAM system evolves stochastically, we may speak of* deterministic *CRN-TAM systems: systems for which there is at most one possible forward reaction, and at most one possible reverse reaction, for every state. That is, the system state space is a one-dimensional line.*

Definition 9. *A CRN-TAM program acting on an initial state L* stops *if the set of reachable states is finite and reaches a state with no possible further reactions with probability one.*

Definition 10. *A CRN-TAM program* constructs *a shape S if the program stops with precisely one terminal assembly, and that assembly has shape S.*

To give a natural notion of the "size" of a CRN-TAM program, we introduce a notion of program complexity. This notion is analogous to the tile set size under the aTAM, or the number of signals and reactions in a CRN.

Definition 11. *The* complexity *of an initial state $I : (S \cup T) \to \mathbb{N}$ is*

$$|I| = \sum_{z \in (S \cup T)} \log_2(I(z) + 1)$$

This definition is natural since it is the number of bits needed to specify a general initial state I, up to small constant multiplicative and additive factors.

Definition 12. *Let $P = (S, T, R, \tau, I)$ be a CRN-TAM program (with unit reaction rate constants). The* complexity *of P with respect to temperature τ is*

$$K_{\mathrm{CT}}^\tau(P) = |S| + |T| + |R| + |I| = |S| + |T| + |R| + \sum_{z \in (S \cup T)} \log_2(I(z) + 1)$$

Each of the terms is related, up to logarithmic factors, to the amount of information needed to specify that component of a CRN-TAM program.

3   Preliminary Results

The CRN-TAM is based on its eponymous models: abstract chemical reaction networks and the abstract tile assembly model. As one would hope, it subsumes both of these models. Subsuming models of stochastic CRNs is trivial:

Theorem 1. *For any chemical reaction network C with species S and reactions R (each of which has at most two reactants and two products), there is a CRN-TAM program $P = (S, \varnothing, R, 0, L)$ with dynamics identical to those of C acting on L.*

In contrast, the abstract tile assembly model requires an unbounded supply of each tile type. Thankfully, it is straightforward to generate this supply with a CRN-TAM program:

Theorem 2. *Let T be a set of tiles for the abstract tile assembly model at temperature τ, and suppose that $\boxed{T_0} \in T$ is the designated seed tile. There is a CRN-TAM program P that simulates the operation of T, in terms of reachable assemblies, with complexity $K_{\mathrm{CT}}^{\tau}(P) \in \Theta(|T|)$.*

Proof. For each tile $\boxed{t} \in T$, we introduce the species C_t and the catalytic reaction $C_t \to C_t + \boxed{t}$. The removal signal of every tile is ε to enforce irreversibility of tile addition. Our initial state consists of one of each C_t and the seed tile $\boxed{T_0}$. The reaction $\boxed{T_0} \to \boxed{\boxed{T_0}}$ initiates the assembly process. $\qquad\square$

Next, we introduce a number of basic constructions that demonstrate the flavor of CRN-TAM programs. The most important of these gives an efficient way to run a broad-class of CRN-TAM programs a certain number of times.

Definition 13. *A CRN-TAM program C is a* handshake subroutine *with respect to a set of data molecules D if it satisfies:*

Data-Inertness Property: *In any state consisting only of molecules in D, no reaction may occur.*

Single-Entry Property: *There is a species S that initiates the operation of C. That is, no reaction will take place until a single molecule of species S appears, and that molecule is consumed in the first reaction of C.*

Single-Exit Property: *There is a species F that signifies the completion of C's operation; we say that F terminates C. That is, F does not appear while reactions of C are still possible, and F is produced by the last reaction of C that can happen.*

Intuitively, handshake subroutines are programs that we can choose to start and can know have stopped. The definition does not require that C always stops, but in typical usage there will be an argument that it does.

Lemma 1. *Let k be a nonnegative integer and P be a handshake subroutine that is initiated by species P_S and terminated by species P_F. There exists a handshake subroutine* powerCounter(k, A, B, P), *initiated by A and terminated by B, with $\Theta(k)$ additional chemical species and $\Theta(k)$ additional chemical reactions that runs P exactly $2^k - 1$ times.*

Proof. Let $\#(a)$ be the number of molecules of species a at a specified time.

We introduce the sequences of species $X_0, X_1, \ldots, X_{k-1}$, $S_1, \ldots, S_k$, and $Y_0, Y_1, \ldots, Y_{k-1}$, all of which do not appear in P (i.e. are "new" species). We construct our counting circuit as a binary counter, where the pair (X_i, Y_i) is a dual-rail representation of the state of the ith bit of the counter; that is, if $\#(X_i) = 1$, then $\#(Y_i) = 0$, and if $\#(X_i) = 0$, then $\#(Y_i) = 1$. By convention, the value of $\#(X_i)$ is the value of the bit. The S_i species will serve as (single-rail) digit carry markers. For k-bit counting, we produce a new program as follows:

1. For each bit i, we introduce the reactions

$$S_i + Y_i \to X_i + P_S, \quad S_i + X_i \to Y_i + S_{i+1}$$

2. Add the reaction $P_F \to S_0$, to continue counting after C runs once.
3. Add the reaction $A \to P_S$ to start the binary counting.
4. Add the reaction $S_k \to B$ to indicate that the counting is finished.

The initial state of our binary counting program is the full collection of species $Y_0, \ldots, Y_{k-1}$, one copy each, indicating that the counter starts at 0.

Notice that the structure of the reactions ensures that the following properties hold by simple induction:

- At every time between the consumption of A and the creation of B, there is exactly one of the S_i carry species at all times, since every other reaction consumes one of these and produces one of these.
- At any time, exactly one of $\{X_i, Y_i\}$ is present, since every reaction "flips a bit" by consuming one $\{X_i, Y_i\}$ and producing the other.
- The reactions implement precisely the carry behavior of a binary counter with k bits.
- At any time, there is only one reaction that can take place, by the above properties, and so the program works deterministically.
- The initiation signal P_S is released and consumed precisely once for every one of the $2^k - 1$ values that the counter's species' can encode.
- Between successive releases of S_0, the program P is run exactly once.

Observe that in our constructed k-bit counter, we introduce $\Theta(k)$ species and $\Theta(k)$ reactions beyond those of the original program. Thus, the program—which we call powerCounter(k, A, B, P)—has the desired properties.   □

We can easily modify this construction to run a handshake subroutine exactly $n \in \mathbb{N}$ times, using $\Theta(\log n)$ signals and reactions:

Theorem 3. *Let n be a positive integer and P be a handshake subroutine that is initiated by species P_S and terminated by species P_F. There exists a handshake subroutine* binaryCounter(n, A, B, P), *initiated by A and terminated by B, with $\Theta(\log n)$ additional species and $\Theta(\log n)$ additional bimolecular reactions that runs P exactly n times.*

Proof. By Lemma 1, we can construct a handshake subroutine that runs P a total of $2^{\lceil \log n \rceil} - 1$ times using $\Theta(\lceil \log n \rceil) = \Theta(\log n)$ additional species and reactions. Extending this to run P one more time using the reactions $S_{\lceil \log n \rceil} \to P_S$ and $P_F \to B$ instead of $S_{\lceil \log n \rceil} \to B$, we have a handshake subroutine that runs P a total of $2^{\lceil \log n \rceil}$ times.

We further modify our construction from Lemma 1 by adding a different initial state. Let $b_0 b_1 \cdots b_{\lceil \log n \rceil}$ be the unique binary representation of $2^{\lceil \log n \rceil} - n \geq 0$, by definition. Then, our initial state (instead of a full sequence of "off" bits Y_i) will be, for all i, $\#(X_i) = b_i$ and $\#(Y_i) = 1 - b_i$.

In effect, our $\lceil \log n \rceil$-bit counter starts with a value of $2^{\lceil \log n \rceil} - n$, which we identify as "zero", and ends in the state $2^{\lceil \log n \rceil}$, which is therefore identified as $2^{\lceil \log n \rceil} - 2^{\lceil \log n \rceil} + n = n$. This construction adds only $\Theta(1)$ complexity beyond that from Lemma 1 so our program still has $\Theta(\log n)$ additional complexity. $\square$

It is intuitively useful to consider the initial state I of a CRN-TAM program P. However, the following theorem shows that the addition of the initial state does not provide extra *algorithmic power* over the model with no additional state.

Theorem 4. *Let $P = (S, T, R, \tau, I)$ be any CRN-TAM program that cannot start until some species F is released. We define a special signal Q^* and let our initial state $I' = \{Q^*\}$. There exists a program $P' = (S', T', R', \tau, I')$ with $K_{CT}^\tau(P') \in \Theta(K_{CT}^\tau(P))$ that has the same graph of possible states after $\Theta(|I|)$ initial states.*

Proof. Let $Z \subseteq (S \cup T)$ be the set of $s \in (S \cup T)$ with $I(s) > 0$, and let $\tilde{Z} = (z_1, z_2, \ldots, z_{|Z|})$ be an arbitrary ordered sequence of Z.

For each z_i, let C_i be the chemical reaction network $Q_i \to z_i + H_i$, and note that C is a handshake subroutine that simply creates one z_i. We construct P' by augmenting P with signal $Q_{|Z|+1}$, reactions $H_i \to Q_{i+1}$, $Q^* \to Q_1$, and $Q_{|Z|+1} \to F$, and for each z_i, a new signal Q_i and a CRN binaryCounter$(I(z_i), Q_i, H_i, C_i)$ that uses new species each time for its internal operation.

This construction is illustrated in Fig. 2. By Theorem 3, each binary counter will produce precisely $I(z_i)$ of each species z_i when Q_i is present.

We can now show that when Q_{i+1} is released, we have the correct "initial state" counts of all species $z_j, j \leq i$. In the base case, notice that the presence of Q^* caused the release of Q_1, which will cause the release of $I(z_1)$ of species z_1. Now, suppose that when Q_{k+1} is released, we have the correct counts of all species $z_j, j \leq k$. Then, notice that the release of Q_{k+1} initiates the release of exactly $I(z_{k+1})$ of z_{k+1}, and also the release of Q_{k+2}. By induction, when species $Q_{|Z|+1}$ is released, all of the species $z_i \in Z$ will have the correct initial counts.

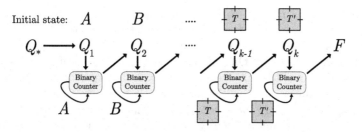

Fig. 2. Conceptual illustration of the construction for Theorem 4. Starting with the singleton "starter species", we use the binary-counting CRNs from Theorem 3 to produce the correct number of each initial species (as indicated above each stage). When each binary counter finishes, it starts the binary counter for the next species.

Since C_i requires a constant number of species and reactions, by Theorem 3, each binary counter and the associated reactions contribute $\Theta(\log I(z_i))$ complexity. Thus, by construction, P' has complexity:

$$K_{CT}^\tau(P') = |S| + |\{Q_i\}_{i=1}^k| + |T| + |R| + \Theta(1) + \sum_{j=1}^k \Theta(\log I(z_j))$$

$$= |S| + |T| + |R| + \sum_{j=1}^k \Theta(\log I(z_j)) = \Theta(K_{CT}^\tau(P)) \qquad \square$$

So long as our CRN-TAM program is deterministic at its starting state, the initial state does not enable asymptotic reduction in program complexity. A careful reader will notice that the notion of program equivalence introduced in Theorem 4 is a restriction of the notion of weak bisimulation, which rigorously establishes a notion of equivalence for concurrent systems. In this paper, all of our constructions use deterministic CRN-TAM programs, and so this theorem will always apply.

This theorem also has a very simple corollary that immediately tells us that the CRN-TAM is in some ways more powerful than the aTAM:

Corollary 1. *A $1 \times n$ rectangle can be constructed by a program P with only a singleton initial state with $K_{CT}^\tau(P) = \mathcal{O}(\log n)$.*

Proof. Let $P = \left(\{S\}, \left\{ \left(\boxed{x}, \varepsilon \right) \right\}, \left\{ S + \boxed{x} \to \boxed{}\boxed{x} \right\}, 1, I \right)$ where $I\left(\boxed{x}\right) = n$, $I(S) = 1$, and $\boxed{x}$ is simply a tile with the same strength-1 glue on two opposite sides (say, east and west). Clearly, P assembles a $1 \times n$ rectangle. By definition, $K_{CT}^\tau(P) = \mathcal{O}(\log n)$, and note that this program will not start building until a seed tile is release. By Theorem 4, there exists a program P' that assembles the $1 \times n$ rectangle using only a singleton initial state with $K_{CT}^\tau(P') = \mathcal{O}(\log n)$. $\square$

In contrast, the size of a tile set (the notion of program complexity for the aTAM) that produces a $1 \times n$ rectangle is $\Theta(n)$ [2]. Similar bounds have been established for other models of tile self-assembly, including the negative glues model [17]. In this example, the lower complexity achieved by the CRN-TAM comes from the explicit control over the number of tiles of a particular type that are present in solution. One can imagine a formulation of the aTAM where similar exact counts of tiles are tracked as they are consumed; in such a model, a program with an initial state consisting of exactly n copies of tile x will construct a $1 \times n$ rectangle with $\mathcal{O}(1)$ tile types. However, Theorem 4 shows that the CRN-TAM can generate its initial state without changing the program complexity, while the analogous result for the modified aTAM would not hold (if program complexity is still taken to be just the number of tile types).

The CRN-TAM also permits the construction of exactly m copies of a shape that can be constructed deterministically:

Theorem 5. *Given a deterministic CRN-TAM program P at temperature τ that constructs a shape $\mathcal{S}$, there is a CRN-TAM program P' that constructs m copies of $\mathcal{S}$ with complexity $K_{CT}^\tau(P') = \mathcal{O}(K_{CT}^\tau(P) + \log m)$.*

Proof (sketch). Since P constructs $\mathcal{S}$ deterministically, at each time there is at most one possible tile addition. Furthermore, the tile addition must be done with a handshake (e.g. $X \to \boxed{t} + W$, $W + t^* \to Y$) because otherwise whether the assembly step or the next reaction occurs first would be non-deterministic. We construct P' by replacing the release and handshake of a single tile by P with a binary counter that, m times, releases the tile and waits for it to attach before continuing. That is, we invoke binaryCounter$(m, X, Y, \{S \to \boxed{t} + W, W + t^* \to F\})$. Note that the first release tile must be able to attach in a unique location on the m identical assemblies, because otherwise P would not have been deterministic. To ensure that each subsequent released tile attaches to a distinct assembly, rather than two or more of them attaching to each other on the same assembly, we label all of the tiles in P with the color red, and introduce an identical set of tiles with color black. We adjust the bonds so that red tiles may only bond to black tiles, and black tiles may only bond to red tiles. At each step, depending on the color of the tile we are trying to bind to, we release a tile of the appropriate color. This creates a checkerboard, and ensures proper assembly. Since P is deterministic, we only need a constant number of binary counters, one for each tile type. By Theorem 3, this takes $\mathcal{O}(\log m)$ extra complexity. Similarly, each step of P that creates a new seed assembly must instead create m seed assemblies. $\qquad\square$

In defining the model, we introduced the condition that a removal reaction may only occur when the corresponding tile is bound to the rest of the assembly with strength exactly the temperature' τ. The next theorem shows that this prevents assemblies from "falling apart" once they have been constructed, except in an order that is approximately the reverse of the order of addition.

Definition 14. *A site (i, j) containing tile $\boxed{x}$ is dependent on site (i', j') containing tile $\boxed{x'}$ if $\boxed{x}$ was added to the assembly after $\boxed{x'}$ and either shares a bond with $\boxed{x'}$ or shares a bond with a tile that is dependent on $\boxed{x'}$.*

This recursive definition of dependency imposes an implicit directed, acyclic graph of dependencies. To disassemble and assembly, we must recursively remove tiles from the leaves of the dependency DAG:

Theorem 6. *If Q is the set of all sites that are dependent on a site (a, b), then for any site $(i, j) \in Q$, the tile at site (i, j) cannot be removed until the tiles at all other sites in Q have been removed. That is, a tile at a site cannot be removed until the tiles at all dependent sites have been removed.*

One (limiting) consequence of Theorem 6 is the impossibility of creating "temporary scaffolding": a CRN-TAM program cannot build permanent parts of an assembly that are dependent on parts that are to be removed later, or it will be impossible to remove the scaffolding.

4   Turing-Universality

With these preliminaries, we consider the ability of CRN-TAM programs to simulate the operation of a Turing machine—which allows it to perform arbitrary computation—under various circumstances.

Theorem 7. *The CRN-TAM is Turing-universal at temperature $\tau = 2$.*

Proof. Given an aTAM tile set T, we may construct a CRN-TAM program $P = (\varnothing, T, \varnothing, 2, I_{\text{seed}})$ that simulates it at temperature 2. Thus, the CRN-TAM is Turing-universal by the Turing-universality of the aTAM [22]. ☐

Since the CRN-TAM subsumes the aTAM, Theorem 7 is far from surprising. However, the construction used to show the Turing-universality of the aTAM relies critically on the ability for Wang tilings to represent computation histories: the state of the Turing machine tape at each step. Importantly, this requires potentially unbounded space in both spatial dimensions. Through clever use of DNA strand displacement polymers, Qian et al. [19] showed the Turing universality of polymer reaction networks by constructing multi-stack machines. Since each stack is one-dimensional, the construction requires unbounded space in only one spatial dimension to provide Turing universality. A related construction for the CRN-TAM provides an analogous result:

Lemma 2. *Consider a (deterministic or non-deterministic) stack machine $M = (Q, \Sigma, \delta, n, q_0)$, consisting of a finite set of states Q, a symbol alphabet Σ, an integer n giving the number of stacks, an initial state $q_0 \in Q$, and a set of transition rules δ where each element of δ is one of:*

1. *$\alpha_1 \rightarrow \alpha_2$ for states $\alpha_1, \alpha_2 \in Q$.*
2. *$\alpha_1 \xrightarrow{\text{pop}_j = \sigma} \alpha_2, \sigma \in \Sigma$ for states $\alpha_1, \alpha_2 \in Q$, corresponding to popping a symbol off of stack j.*
3. *$\alpha_1 \xrightarrow{\text{push}_j(\sigma)} \alpha_2$ for states $\alpha_1, \alpha_2 \in Q$ and symbol $\sigma \in \Sigma$, corresponding to pushing a symbol σ onto stack j.*

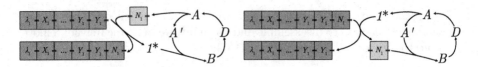

Fig. 3. Conceptual illustrations of the push and pop operations for a stack machine. *On the left*, the push operation "state A goes to state B, pushing symbol N onto stack 1" is shown. During a push operation, a stack-specific symbol is released, along with an intermediate species A'. The finite control waits for the stack- (but not symbol-) specific removal signal 1^* before continuing. *On the right*, the pop operation "from state A, pop a symbol from stack 1. If it is an N, go to state B" is shown. The pop operation is basically the reverse of the push operation. In both diagrams, D represents a network of other reactions, not just a single signal.

There is a CRN-TAM program $P = (S, T, R, 1, I)$ that simulates M with $K^1_{CT}(P) = \mathcal{O}(|Q| + |\Sigma| + |\delta|)$. Furthermore, it requires unbounded space in only one geometric dimension, and runs in constant space in the other.

Proof. We show the result by construction, showing how to "compile" a stack machine M into a CRN-TAM program P. Our construction, based on the one used by [19] and illustrated in Fig. 3, is:

- For each state $\alpha_i \in Q$, we introduce the species S_i and S'_i, including the initial state q_0 (represented by S_0).
- For each stack k, we introduce:
  - A "stack query species" Q_k
  - For each symbol $\sigma \in \Sigma$, a tile $\boxed{\sigma_k} = (\varnothing, -k, \varnothing, -k)$, each with the same removal signal Q_k
  - A tile $\boxed{\lambda_k} = (\varnothing, -k, \varnothing, \varnothing)$ that represents the "bottom-of-stack", with removal signal ε.
- For each transition in δ:
  1. Implement transitions of the form $\alpha_i \to \alpha_j$ with the reaction

  $$S_i \to S_j$$

  2. Implement transitions of the form $\alpha_i \xrightarrow{\text{pop}_\ell = \sigma} \alpha_j$ with the reactions

  $$S_i \to Q_\ell + S'_i, \quad \boxed{\sigma_\ell} + S'_i \to S_j$$

  3. Implement transitions of the form $\alpha_i \xrightarrow{\text{push}_\ell(\sigma)} \alpha_j$ with the reactions

  $$S_i \to \boxed{\sigma_\ell} + S'_i, \quad S'_i + Q_\ell \to S_j$$

- Include as initial state one of each $\boxed{\lambda_k}$ and one of species S_0. Include the reactions $\boxed{\lambda_k} \to \boxed{\boxed{\lambda_k}}$ to produce the initial assemblies.

The idea behind this construction is to represent the finite state with the presence of one of the S_i signals, and use tiles to construct a physical "stack" assembly of tiles for each stack in the stack machine. By identifying each tile with a stack's identity, we are able to restrict the symbol to interactions with the desired stack. To show correctness, we demonstrate that the transitions are correct, and argue that the possible reactions at each point are precisely the valid transitions from the current state in δ.

Clearly, reactions of type (1) are correct, since the reaction $S_i \rightarrow S_j$ is a direct implementation of the transition $\alpha_i \rightarrow \alpha_j$.

Reactions that perform pushes and pops are somewhat more complicated. To pop a symbol from stack ℓ, we produce the stack query species Q_ℓ and the "state storage species" S_i' that indicates that we are in the process of transitioning out of state α_i. Since Q_ℓ is the removal species of every one of the $\boxed{\sigma_\ell}, \forall \sigma \in \Sigma$, this will remove the tile at the only "exposed" site of stack ℓ. Then, this tile $\boxed{\sigma_\ell}$ can react with S_i' to produce the state transition to S_j.

Similarly, to push symbol σ onto stack ℓ, we release $\boxed{\sigma_\ell}$ with the reaction $S_i \rightarrow \boxed{\sigma_\ell} + S_i'$, releasing S_i' to indicate that we are in the process of transitioning out of state α_i. When the tile $\boxed{\sigma_\ell}$ attaches to the assembly for stack ℓ, it will release its removal species Q_ℓ. The removal species will then react with S_i', which transitions to another state.

Lastly, note that at any point in time, exactly one of the S_i or S_i' is present, and every reaction both produces and consumes exactly one S_i or S_i'. Thus, the progress of the CRN is deterministic, except possibly where a single S_i could transition in several ways (if the original stack machine is non-deterministic). We introduce a constant number of species and reaction to represent each state, stack symbol, and transition rule, and so $K_{CT}^1(P) = \mathcal{O}(|Q| + |\Sigma| + |\delta|)$.   □

Theorem 8. *Let U be a Turing machine. There exists a CRN-TAM program that simulates the operation of U, and requires unbounded space in only one geometric dimension, while running in constant space in the other. That is, the CRN-TAM is Turing-universal when running in one spatial dimension.*

Proof. It is well known that multi-stack pushdown automata are equivalent to Turing machines [26]. Observe that a multi-stack pushdown automaton can be implemented using the same construction used in Lemma 2. Thus, U may be implemented by way of an equivalent multi-stack pushdown automaton.   □

There are two critical differences between Theorems 7 and 8. First, the aTAM is Turing-universal only at temperature 2, since algorithmic self-assembly is required for universality. In contrast, the CRN-TAM can simulate stack machines at temperature 1, and is thus Turing-universal at all nonzero temperatures. Although this is believed to be impossible for the temperature 1 aTAM, a 3D generalization aTAM is Turing-universal at temperature 1; furthermore, the construction requires only constant space in the third dimension [9].

Second, the CRN-TAM supports Turing-universal computation using unbounded space in only one spatial dimension, while the aTAM requires unbounded space in both spatial dimensions for universality. Interestingly, negative glues also allow for a restricted version of this result, where assemblies can be split into one-dimensional assemblies but cannot be deconstructed completely [12].

5 Optimal Encoding of Binary Strings

Having given efficient constructions for simulating Turing-universal computation, we now consider the related problem of efficiently encoding inputs for these Turing machines as (binary) strings. Given our stack machine construction, we aim to *encode* strings of length n as $1 \times n$ tile assemblies, which can be used as the input for a stack machine like in Lemma 2.

Definition 15. *A CRN-TAM program P encodes a binary string x if it constructs a $1 \times |x|$ rectangular assembly of tiles representing the bits of x.*

Of course, a binary string x of length n can be easily encoded by a CRN-TAM program of $\Theta(n)$ unique tile types: one for each bit of x. However, just as Adleman et al. [1] and Soloveichik and Winfree [29] encoded strings of length n in smaller aTAM tile sets that self-assemble at temperature $\tau = 2$ to unpack the bits, in the CRN-TAM we can do substantially better at $\tau = 1$ and using just one dimension for self-assembly:

Theorem 9. *For any binary string x of length n, there is a CRN-TAM program $P = (S, T, R, 1, I)$ that encodes x and has complexity $K^1_{CT}(P) = \mathcal{O}(n/\log n)$.*

Proof. Suppose that we represent x as a sequence of k binary words $(w_1, w_2, \ldots, w_k)$, each with size $w = n/k$. For convenience, define the functions $h(x)$ and $t(x)$ to be the head and tail of a binary string x (i.e. the first bit, and all the remaining bits, respectively).

Define the data tiles $\boxed{T_0} = (\varnothing, -, \varnothing, -)$ and $\boxed{T_1} = (\varnothing, -, \varnothing, -)$, encoding the binary symbols zero and one, with removal signals 0^* and 1^*, respectively. Additionally, define the "bottom-of-stack" tile $\boxed{\lambda} = (\varnothing, -, \varnothing, \varnothing)$ that will create the seed assembly through an activation reaction.

For every binary string a of length at most w, we introduce several signals: A_a (the "construction signal"), A'_a (the "intermediate signal"), W_a (the "wait signal"), and B_a (the "completion signal"). For any binary string a of length at most w, define the following set R_a of reactions:

$$R_a = \{A_a \rightarrow A'_a + \boxed{T_{h(w)}}, \quad h(w)^* + A'_a \rightarrow W_a + A_{t(w)}, \quad B_{t(w)} + W_a \rightarrow B_w\}$$

By construction, the reactions in R_a push the first bit of a onto the stack, then invoke the reaction gadget for the remaining bits of a. It waits to receive the completion signal for the tail bits of a, and then issues its own completion signal. Notice that R_a has constant size.

To encode x, we can use k hard-coded tiles of distinct tile types that will assemble together; each tile represents a w-bit word of x; equivalently, we could use a hard-coded CRN producing unique signals for each word. We can then include the tiles, signals, and reactions described above for every string of length at most w, along with reactions that will repeatedly pop a word-tile representing word a from the stack of word tiles, produce A_a, and wait to consume B_a before popping the next word tile. These require only a constant number of reactions, so the total program complexity is: $K_{CT}^1(P) = \Theta(k) + \mathcal{O}(2^{w+1})$ since there are 2^{w+1} strings of length at most w.

Picking $w = \log(n/\log n)$, we get $k = n/(\log n - \log \log n) \in \mathcal{O}(n/\log n)$, and so $K_{CT}^\tau(P) = \mathcal{O}(n/\log n)$. $\qquad\square$

6   Kolmogorov-Optimal Assembly of Algorithmic Shapes

We now turn our attention to the problem of constructing a geometric shape $\mathcal{S}$ using the CRN-TAM program of minimal complexity. Following [29], we can give a formal definition of shape:

Definition 16. *A shape $\mathcal{S}$ is a connected subset of the 2-dimensional lattice $\mathbb{Z}^2$ under the equivalence relation $\mathcal{S} = \mathcal{S}'$ if and only if $\mathcal{S}'$ is a translation of $\mathcal{S}$. A c-scaling of a shape $\mathcal{S}$ is the shape $s_c(\mathcal{S})$ that is obtained by replacing each square of $\mathcal{S}$ with a $c \times c$ block of squares.*

We follow the usual definition of the *Kolmogorov complexity* of a binary string x with respect to a fixed universal Turing machine U as the minimal size of a program for U that outputs x. We extend this notion to a shape $\mathcal{S}$:

Definition 17. *The Kolmogorov complexity of a shape $\mathcal{S}$ with respect to a universal Turing machine U is the minimal size of a program for U that outputs $\mathcal{S}$ as a list of coordinates. We denote the Kolmogorov complexity of $\mathcal{S}$ by $K(\mathcal{S})$.*

Definition 18. *The CRN-TAM complexity of a shape $\mathcal{S}$ at temperature τ is the minimum complexity of any CRN-TAM program that constructs $\mathcal{S}$. We denote the CRN-TAM complexity of $\mathcal{S}$ at temperature τ as $K_{CT}^\tau(\mathcal{S})$.*

Notice that since we may efficiently simulate low-temperature programs at higher temperatures, $K_{CT}^\tau(\mathcal{S}) \geq K_{CT}^{\tau+1}(\mathcal{S})$.

Theorem 10. *For any shape $\mathcal{S}$, the CRN-TAM complexity of $s_2(\mathcal{S})$ at any temperature $\tau \geq 1$ satisfies:*

$$K_{CT}^\tau(s_2(\mathcal{S})) \in \Theta\left(\frac{K(\mathcal{S})}{\log K(\mathcal{S})}\right)$$

We will prove this theorem as two lemmas: Lemma 3 for the upper bound, and Lemma 4 for the lower bound.

Lemma 3. *For any shape S, there is a CRN-TAM program $P = (S, T, R, 1, I)$ with*

$$K^1_{CT}(P) \in \mathcal{O}\left(\frac{K(S)}{\log K(S)}\right)$$

that constructs S at scale 2.

Proof. Inspired by Soloveichik and Winfree [29], our proof gives a construction of a CRN-TAM program that satisfies the complexity bound. Our construction uses a simulated Turing machine and a finite set of "path building" tiles.

In our construction, we reduce the problem of constructing S at scale 2 to the problem of constructing a path around a spanning tree of S. To do this, we use a set of tiles T_c that consists of all possible tiles with exactly two strength-1 bonds with the same label. These tiles can produce any path in the 2D lattice; the specific path is determined by the sequence in which tiles are released.

Let U be a universal Turing machine, and define ψ to be a program for U that outputs the $\mathbb{Z}^2$ coordinates of each point in S. We construct a program φ for U that does the following:

1. Run ψ to obtain the lattice points in S.
2. From some point in S, use a depth-first search to find a spanning tree of S.
3. Construct a path $W \subseteq \mathbb{Z}^2$ through the (scale 2) lattice that walks around the perimeter of the spanning tree.

Our CRN-TAM program uses the construction from Theorem 8 to simulate U running φ. We may assume without loss of generality that U acts on a binary alphabet. Then, using the implementation of U, the CRN-TAM program begins at the start of path W and releases appropriate tiles one-by-one to fill in the lattice sites occupied by W. This construction process is illustrated in Fig. 4.

Since every shape has a spanning tree, we may always find one with depth-first search. Furthermore, notice that when we scale the spanning tree to scale 2, there is always space for a perimeter walk W. Thus, P will construct $s_2(S)$.

Now, suppose that ψ is a Kolmogorov-optimal Turing machine program that outputs S, so that $|\psi| = K(S)$. Since all parts of φ other than ψ are independent of the shape S and thus constant, $|\varphi| = \Theta(|\psi|) = \Theta(K(S))$. By encoding φ using the optimal encoding construction in Theorem 9, which works at temperature 1, we can encode and unpack φ in $\mathcal{O}(|\varphi|/\log|\varphi|) = \mathcal{O}(K(S)/\log K(S))$ CRN-TAM program complexity. Using the construction from Theorem 8, we may simulate the universal Turing machine U with constant program complexity. The complexity of P is the sum of the complexity of the universal Turing machine and the encoding of the optimal program, so

$$K^1_{CT}(P) = \mathcal{O}(1) + \mathcal{O}\left(\frac{K(S)}{\log K(S)}\right) = \mathcal{O}\left(\frac{K(S)}{\log K(S)}\right) \qquad \square$$

Lemma 4. *For any shape S, every CRN-TAM program $P = (S, T, R, \tau, I)$ that constructs S at fixed scale m has $K^\tau_{CT}(P) \log K^\tau_{CT}(P) \in \Omega(K(S))$.*

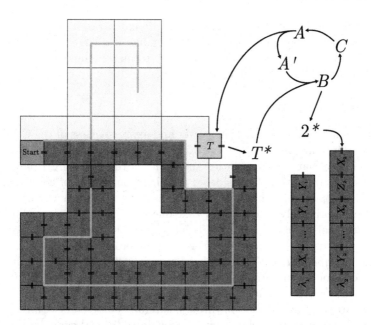

Fig. 4. Conceptual illustration of the construction of a shape $\mathcal{S}$ at scale 2 using a Kolmogorov-optimal CRN-TAM program.

Proof. First, note that there is a constant size Turing machine program p_{sim} that takes a binary description of a CRN-TAM program and simulates its operation by traversing the reachability graph of states. We construct our program so that it will output the coordinates of the occupied squares of the final assembly if the program constructs a shape. If the program stops without meeting this condition, it indicates failure.

By efficiently encoding the signals, tiles, reactions, and initial state in binary, we can represent a CRN-TAM program $P = (S, T, R, \tau, I)$ as input to our simulator in $\mathcal{O}(K_{\text{CT}}^{\tau}(P) \log K_{\text{CT}}^{\tau}(P))$ bits. By definition,

$$K(\mathcal{S}) \leq |p_{\text{sim}}| + \mathcal{O}(K_{\text{CT}}^{\tau}(P) \log K_{\text{CT}}^{\tau}(P)) = \mathcal{O}(K_{\text{CT}}^{\tau}(P) \log K_{\text{CT}}^{\tau}(P))$$

$\square$

With Theorem 10, we demonstrate the algorithmic power of the CRN-TAM over previous models of tile-based self-assembly. Although an analogous result holds for the aTAM, it allows construction of a shape $\mathcal{S}$ only at a (possibly very) large scale c, which is polynomial in the runtime of U on ψ [29]. Constant-scale construction of algorithmic shapes with Kolmogorov-optimal tile sets is possible with temperature programming; however, these results require a number of temperature changes that is linear in the size of the shape [30]. The number of temperature changes should be a part of the natural definition of program complexity for temperature programming models. In contrast, the CRN-TAM permits construction of shapes at scale 2 with a Kolmogorov-optimal program complexity.

In the full version of this paper, we will present a result that extends this construction to give Kolmogorov-optimal assembly of a large class of shapes at scale one. It remains open to show that all shapes can (or cannot) be constructed by Kolmogorov-optimal CRN-TAM programs.

7   Open Questions

Although we have demonstrated the power and expressiveness of the CRN-TAM for Turing-universal computation and Kolmogorov-optimal construction, many open questions about the capabilities and limits of the CRN-TAM remain.

In our work, we have not considered the time complexity of computation or construction. In fact, many of our constructions proceed quite slowly under Gillespie dynamics, primarily because they take one step at a time—according to the rather limited notion of deterministic behavior used here to make our constructions simple to analyze. There are therefore numerous open questions related to the time complexity of CRN-TAM programs, such as how fast a shape can be constructed or a computation can be performed. Chemistry is an inherently parallel computational medium, yet all of our constructions have been designed to engineer around this parallelism through carefully enforced determinism. How to exploit the parallelism of chemistry to provide additional expressive power in the CRN-TAM remains to be seen.

Lastly, an important task is to develop molecular motifs that can implement CRN-TAM programs. While designs may build on the work by Zhang et al. [33], in which a DNA strand displacement circuit controlled the activation of DNA double-crossover tiles, there is a substantial difficulty with implementing the CRN-TAM based on the crystal-growth mechanism inherent in simple tile self-assembly: because (unlike in the aTAM) our model does not hold tile concentrations constant, we cannot justify low-error rates based on assuming that growth occurs near the (concentration-dependent) melting temperature for attachment by τ bonds. A more suitable molecular implementation might be based on components that become activated for further assembly by some configurational change, such as the elegant one-dimensional hybridization chain reaction [10] as generalized for a restricted class of signal tiles [15]. Such mechanisms are well suited to $\tau = 1$ seeded assembly, but have not yet been generalized for two dimensional assembly or for $\tau = 2$. We expect that the chief difficulty in a molecular implementation of CRN-TAM programs will be enforcing the proper interactions between tiles and their removal signals. However, previous work on implementing stack machines with DNA strand displacement reactions [19] proposed an implementation of a similar mechanism for handshaking assembly steps when constructing one-dimensional assemblies. Known implementations for many CRN-TAM features plausibly suggest the existence of a physical implementation of the CRN-TAM.

Acknowledgements. We acknowledge financial support from National Science Foundation grant CCF-1317694. We also thank Dave Doty, for his helpful comments and suggestions, and Kevin Li, for his useful suggestions on an early draft of this paper.

References

1. Adleman, L., Cheng, Q., Goel, A., Huang, M.D.: Running time and program size for self-assembled squares. In: ACM Symposium on Theory of Computing (STOC), pp. 740–748 (2001)
2. Aggarwal, G., Cheng, Q., Goldwasser, M.H., Kao, M.Y., de Espanes, P.M., Schweller, R.T.: Complexities for generalized models of self-assembly. SIAM J. Comput. **34**(6), 1493–1515 (2005)
3. Barish, R.D., Schulman, R., Rothemund, P.W., Winfree, E.: An information-bearing seed for nucleating algorithmic self-assembly. Proc. Natl. Acad. Sci. **106**(15), 6054–6059 (2009)
4. Bennett, C.H.: The thermodynamics of computation - a review. Int. J. Theor. Phys. **21**(12), 905–940 (1982)
5. Cardelli, L., Zavattaro, G.: On the computational power of biochemistry. In: Horimoto, K., Regensburger, G., Rosenkranz, M., Yoshida, H. (eds.) AB 2008. LNCS, vol. 5147, pp. 65–80. Springer, Heidelberg (2008)
6. Chen, H.L., Doty, D., Soloveichik, D.: Deterministic function computation with chemical reaction networks. Nat. Comput. **13**(4), 517–534 (2014)
7. Chen, Y.J., Dalchau, N., Srinivas, N., Phillips, A., Cardelli, L., Soloveichik, D., Seelig, G.: Programmable chemical controllers made from DNA. Nat. Nanotechnol. **8**(10), 755–762 (2013)
8. Condon, A., Hu, A.J., Maňuch, J., Thachuk, C.: Less haste, less waste: on recycling and its limits in strand displacement systems. Interface Focus **2**(4), 512–521 (2012)
9. Cook, M., Fu, Y., Schweller, R.: Temperature 1 self-assembly: deterministic assembly in 3D and probabilistic assembly in 2D. In: ACM-SIAM Symposium on Discrete Algorithms (SODA), pp. 570–589. SIAM (2011)
10. Dirks, R.M., Pierce, N.A.: Triggered amplification by hybridization chain reaction. Proc. Natl. Acad. Sci. **101**(43), 15275–15278 (2004)
11. Doty, D.: Theory of algorithmic self-assembly. Commun. ACM **55**(12), 78–88 (2012)
12. Doty, D., Kari, L., Masson, B.: Negative interactions in irreversible self-assembly. Algorithmica **66**, 153–172 (2013)
13. Gillespie, D.T.: A general method for numerically simulating the stochastic time evolution of coupled chemical reactions. J. Comput. Phys. **22**(4), 403–434 (1976)
14. Ke, Y., Ong, L.L., Shih, W.M., Yin, P.: Three-dimensional structures self-assembled from DNA bricks. Science **338**(6111), 1177–1183 (2012)
15. Padilla, J.E., Sha, R., Kristiansen, M., Chen, J., Jonoska, N., Seeman, N.C.: A signal-passing DNA-strand-exchange mechanism for active self-assembly of DNA nanostructures. Angew. Chem. Int. Ed. **54**(20), 5939–5942 (2015)
16. Patitz, M.J.: An introduction to tile-based self-assembly and a survey of recent results. Nat. Comput. **13**(2), 195–224 (2013)
17. Patitz, M.J., Schweller, R.T., Summers, S.M.: Exact shapes and turing universality at temperature 1 with a single negative glue. In: Cardelli, L., Shih, W. (eds.) DNA 17 2011. LNCS, vol. 6937, pp. 175–189. Springer, Heidelberg (2011)
18. Pinheiro, A.V., Han, D., Shih, W.M., Yan, H.: Challenges and opportunities for structural DNA nanotechnology. Nat. Nanotechnol. **6**(12), 763–772 (2011)
19. Qian, L., Soloveichik, D., Winfree, E.: Efficient turing-universal computation with DNA polymers. In: Sakakibara, Y., Mi, Y. (eds.) DNA 16 2010. LNCS, vol. 6518, pp. 123–140. Springer, Heidelberg (2011)

20. Qian, L., Winfree, E.: Scaling up digital circuit computation with DNA strand displacement cascades. Science **332**(6034), 1196–1201 (2011)
21. Rothemund, P.W.K., Papadakis, N., Winfree, E.: Algorithmic self-assembly of DNA Sierpinski triangles. PLoS Biol. **2**(12), e424 (2004)
22. Rothemund, P.W.K., Winfree, E.: The program-size complexity of self-assembled squares. In: ACM Symposium on Theory of Computing (STOC), pp. 459–468. ACM (2000)
23. Rothemund, P.W., Ekani-Nkodo, A., Papadakis, N., Kumar, A., Fygenson, D.K., Winfree, E.: Design and characterization of programmable DNA nanotubes. J. Am. Chem. Soc. **126**(50), 16344–16352 (2004)
24. Seelig, G., Soloveichik, D., Zhang, D.Y., Winfree, E.: Enzyme-free nucleic acid logic circuits. Science **314**(5805), 1585–1588 (2006)
25. Seeman, N.C.: An overview of structural DNA nanotechnology. Mol. Biotechnol. **37**(3), 246–257 (2007)
26. Sipser, M.: Introduction to the Theory of Computation. Cengage Learning, Boston (2012)
27. Soloveichik, D., Cook, M., Winfree, E., Bruck, J.: Computation with finite stochastic chemical reaction networks. Nat. Comput. **7**(4), 615–633 (2008)
28. Soloveichik, D., Seelig, G., Winfree, E.: DNA as a universal substrate for chemical kinetics. Proc. Natl. Acad. Sci. **107**(12), 5393–5398 (2010)
29. Soloveichik, D., Winfree, E.: Complexity of self-assembled shapes. SIAM J. Comput. **36**(6), 1544–1569 (2007)
30. Summers, S.M.: Reducing tile complexity for the self-assembly of scaled shapes through temperature programming. Algorithmica **63**(1–2), 117–136 (2011)
31. Wei, B., Dai, M., Yin, P.: Complex shapes self-assembled from single-stranded DNA tiles. Nature **485**(7400), 623–626 (2012)
32. Yin, P., Choi, H.M.T., Calvert, C.R., Pierce, N.A.: Programming biomolecular self-assembly pathways. Nature **451**(7176), 318–322 (2008)
33. Zhang, D.Y., Hariadi, R.F., Choi, H.M.T., Winfree, E.: Integrating DNA strand-displacement circuitry with DNA tile self-assembly. Nat. Commun. **4** (2013). Article No. 1965
34. Zhang, D.Y., Seelig, G.: Dynamic DNA nanotechnology using strand-displacement reactions. Nat. Chem. **3**(2), 103–113 (2011)
35. Zhang, D.Y., Turberfield, A.J., Yurke, B., Winfree, E.: Engineering entropy-driven reactions and networks catalyzed by DNA. Science **318**(5853), 1121–1125 (2007)

Reflections on Tiles (in Self-Assembly)

Jacob Hendricks[1], Matthew J. Patitz[1]([✉]), and Trent A. Rogers[2]

[1] Department of Computer Science and Computer Engineering,
University of Arkansas, Fayetteville, USA
{jhendric,patitz}@uark.edu
[2] Department of Mathematical Sciences, University of Arkansas, Fayetteville, USA
tar003@email.uark.edu

Abstract. We define the Reflexive Tile Assembly Model (RTAM), which is obtained from the abstract Tile Assembly Model (aTAM) by allowing tiles to reflect across their horizontal and/or vertical axes. We show that the class of directed temperature-1 RTAM systems is not computationally universal, which is conjectured but unproven for the aTAM, and like the aTAM, the RTAM is computationally universal at temperature 2. We then show that at temperature 1, when starting from a single tile seed, the RTAM is capable of assembling $n \times n$ squares for n odd using only n tile types, but incapable of assembling $n \times n$ squares for n even. Moreover, we show that n is a lower bound on the number of tile types needed to assemble $n \times n$ squares for n odd in the temperature-1 RTAM. The conjectured lower bound for temperature-1 aTAM systems is $2n - 1$. Finally, we give preliminary results toward the classification of which finite connected shapes in $\mathbb{Z}^2$ can be assembled (strictly or weakly) by a singly seeded (i.e. seed of size 1) RTAM system, including a complete classification of which finite connected shapes may be strictly assembled by a *mismatch-free* singly seeded RTAM system.

1 Introduction

Self-assembly is the process by which disorganized components autonomously combine to form organized structures. In DNA-based self-assembly, the combining ability of the components is implemented using complementary strands of DNA as the "glue". In [23], Winfree introduced a useful mathematical model of self-assembling systems called the abstract Tile Assembly Model (aTAM) where the autonomous components are described as square tiles with specifiable glues on their edges and the attachment of these components occurs spontaneously when glues match. The aTAM provides a convenient way of describing self-assembling systems and their resulting assemblies, and serves as the underpinning of many

J. Hendricks and M.J. Patitz—Supported in part by National Science Foundation Grant CCF-1117672 and CCF-1422152.

T.A. Rogers—This author's research was supported by the National Science Foundation Graduate Research Fellowship Program under Grant No. DGE-1450079, and National Science Foundation grants CCF-1117672 and CCF-1422152.

© Springer International Publishing Switzerland 2015
A. Phillips and P. Yin (Eds.): DNA 2015, LNCS 9211, pp. 55–70, 2015.
DOI: 10.1007/978-3-319-21999-8_4

studies of the properties of self-assembling systems. For a comprehensive survey of tile-based self-assembly including models other than the aTAM, see [5,16].

From the broad collection of results in the aTAM, one property of systems that has been shown to yield enormous power is *cooperation*. The notion of cooperation captures the phenomenon where the attachment of a new tile to a growing assembly requires it to bind to more than one tile (usually 2) already in the assembly. The requirement for cooperation is determined by a system parameter known as the *temperature*, and when the temperature is equal to 1 (a.k.a. temperature-1 systems), there is no requirement for cooperation. A long-standing conjecture is that temperature-1 aTAM systems are in fact not capable of universal computation or efficient shape building, although it is well-known that temperature ≥ 2 systems are. However, in actual laboratory implementations of DNA-based tiles [1,15,20,22,24], the self-assembly performed by temperature-2 systems does not match the error-free behavior dictated by the aTAM, but instead, a frequent source of errors is the binding of tiles using only a single bond. Thus, temperature-1 behavior erroneously occurs and cannot be completely prevented.

Many models of self-assembly can be thought of as extensions of the aTAM (e.g. [2,4,6,8,9,17]), and for these models it is common to study the added power that an extra property or constraint gives the extended model. For example, in [2,6,8,17], it is shown that at temperature 1, when the aTAM is appropriately extended, the resulting models are computationally universal and capable of efficiently assembling shapes. In this paper, we take the opposite approach and remove a constraint that the aTAM imposes with the goal of modelling physical systems that may be incapable of enforcing these constraints. Tiles in the aTAM are not allowed to flip or rotate prior to attachment to an existing assembly. While this assumption is a realistic one for many implementations of DNA-based tiles (e.g. [23]), for certain implementations of DNA-based building blocks (e.g. nBLOCKs [13], where separate, disconnected strands of DNA are attached to different sides of nanoparticles), it is unknown whether or not both of the conditions of this assumption can be physically enforced. (See [3,10,12,18] for more experimentally produced building blocks and systems.) When DNA is used as the binding agent, single stranded DNA can be used to prevent relative tile rotation by encoding a direction (north/south or east/west) in the DNA sequence so that only strands with appropriately matching directions are complementary; on the other hand, preventing tiles from flipping may not always be possible, especially if the glues of different sides of a tile are encoded by disjoint DNA complexes. Therefore, we consider a model based on the aTAM where tiles may nondeterministically flip horizontally and/or vertically prior to attachment.

We introduce the *Reflexive Tile Assembly Model* (RTAM), which can be thought of as the aTAM with the relaxed constraint that tiles in the RTAM *are* allowed to flip horizontally and/or vertically. Also, unlike most formulations of the aTAM where complementary strands of DNA are represented with the same glue label, the RTAM explicitly specifies complementary glues. This importantly prevents copies of tiles of the same type from being able to flip and bind to each other, and is the actual reality with DNA-based tiles. (A simple example of an RTAM system can be seen in Fig. 1). We then show a series of results within the RTAM. First we show that at temperature 1, the class of directed RTAM systems – systems which yield a single pattern up to reflection and ignoring tile orientation – are only capable of assembling patterns that are essentially periodic. Then, following the thesis set forth in [7], we conclude that the temperature-1 RTAM is not computationally universal. While the inability of temperature-1 aTAM systems to compute is still only conjectured, we are able to conclusively prove it for RTAM systems, specifically by using techniques developed in [7] to study temperature-1 aTAM systems. We also show that like the aTAM at temperature 2, the class of directed temperature-2 RTAM systems is computationally universal. This shows a fundamental dividing line between the powers of RTAM temperature-1 and temperature-2 directed systems. We then turn our attention to the self-assembly of squares by singly seeded temperature-1 RTAM systems where we show that for even values of $n \in \mathbb{N}$ it is impossible to self-assemble any $n \times n$ square. This is exceptional due to the fact that it is the first demonstration of a model of tile assembly in which a finite shape is proven the be impossible to self-assemble in a directed system. Typically, any finite shape can be self-assembled by a trivial system in which a unique tile type is created for each point of the shape. However, due to the ability of tiles in the RTAM to flip, it is not possible for the RTAM systems to effectively constrain the reflections of tiles to produce such even squares without the possibility of tiles growing beyond the boundaries of the squares. However, for odd values of n and m any $n \times m$ rectangle can be self-assembled using only $\frac{n+m}{2}$ tile types, thus implying that for n odd, an $n \times n$ square can be self-assembled using only n tile types. In addition, we also show that for n odd, an $n \times n$ square cannot be self-assembled using less than n tile types (thus, n is the upper and lower bound for square assembly). This is in contrast to the aTAM at temperature 1, where the conjectured lower bound for assembling an $n \times n$ square is $2n - 1$, and hints that in certain situations the ability of RTAM tiles to attach in flipped orientations can be effectively harnessed to more efficiently build shapes than systems in the aTAM. Finally, we give preliminary results toward the classification of the finite connected shapes in $\mathbb{Z}^2$ that can be assembled (strictly or weakly) by a singly seeded RTAM system, including a complete classification of which finite connected shapes be strictly assembled by a *mismatch-free* singly seeded temperature-1 RTAM system. We also show that arbitrary shapes with scale factor 2 can be assembled in the singly seeded temperature-2 RTAM. These combined results show that the ability of tiles to bind in flipped orientations is sometimes provably limiting, while at other times can provide advantages, and they provide a solid framework for the study of self-assembling systems composed of molecular building blocks unable to enforce the constraints of the aTAM.

The layout of this paper is as follows. In Sect. 2 we present the definition of the RTAM. Section 3 contains the proof that temperature-1 RTAM systems cannot perform universal computation and that temperature-2 systems can. In Sect. 4 we present our results related to the self-assembly of shapes in the RTAM, including our results about assembling squares and classifying the finite connected shapes that self-assemble in the RTAM. (Due to space constraints, full proofs of each result can be found in [11]).

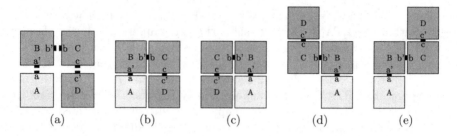

Fig. 1. An example RTAM system. (a) Tile set, (b)–(e) the set of assemblies which could grow from the seed consisting of an A tile. The assemblies of (b) and (c) are considered to be the same assembly since one can be achieved from the other by a horizontal reflection, as are those of (d) and (e). Note that the assembly of (b) is the only assembly that could form in the aTAM, while in the RTAM this system makes two distinct assemblies.

2   Definition of the Reflexive Tile Assembly Model

The Reflexive Tile Assembly Model (RTAM) is essentially equivalent to the abstract Tile Assembly Model (aTAM) [14,19,21,23] but with the modification that tiles are allowed to possibly "flip" across their horizontal and/or vertical axes before attaching to an assembly. Also, as in some formulations of the aTAM, it is assumed that glues bind to complementary versions of themselves (so that two tiles of the same type but flipped relative to each other can't simply bind to each other along the same but reflected side). We now formally define the RTAM. Our notation is similar (and where appropriate, identical) to that of [14].

We work in the 2-dimensional discrete space $\mathbb{Z}^2$. Define the set $U_2 = \{(0,1), (1,0), (0,-1), (-1,0)\}$ to be the set of all *unit vectors* in $\mathbb{Z}^2$. We also sometimes refer to these vectors by their cardinal directions N, E, S, W, respectively. All *graphs* in this paper are undirected. A *grid graph* is a graph $G = (V, E)$ in which $V \subseteq \mathbb{Z}^2$ and every edge $\{\boldsymbol{a}, \boldsymbol{b}\} \in E$ has the property that $\boldsymbol{a} - \boldsymbol{b} \in U_2$.

Intuitively, a tile type t is a unit square that can be translated and flipped across its vertical and/or horizontal axes, but not rotated. This provides each tile type with a pair of North-South (NS) sides and a pair of East-West (EW) sides, such that either side $s \in NS$ may be facing north while the other is facing south (and vice versa for the EW glues). For ease of discussion, however, we will talk about tile types as being defined in fixed orientations, but then allow them to attach to assemblies in possibly flipped orientations. Therefore, we define each t as having a well-defined "side $\boldsymbol{u}$" for each $\boldsymbol{u} \in U_2$. Each side $\boldsymbol{u}$ of t has a "glue" with "label" $label_t(\boldsymbol{u})$–a string over some fixed alphabet–and "strength" $str_t(\boldsymbol{u})$–a nonnegative integer–specified by its type t. Let $R = \{D, V, H, B\}$ be the set of permissible reflections for a tile which is assumed to begin in the default orientation, where D corresponds to no change from the default, V a single vertical flip (i.e. a reflection across the x-axis), H a single horizontal flip (i.e. a reflection across the y-axis), and B a single horizontal flip and a single vertical flip. (Note that the ordering of flips for B does not matter as

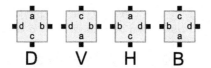

Fig. 2. Left to right: (1) Default orientation of an example tile type t, (2) t flipped vertically, (3) t flipped horizontally, (4) t flipped across both axes

either ordering results in the same orientation, and also that all combinations of possibly many flips across each axis result only in tiles of the orientations provided by R.) See Fig. 2 for an example of each. Let $S : R \times U_2 \rightarrow U_2$ be a function which takes a type of reflection $r \in R$ and a side $s \in U_2$, and which returns the side of a tile in its default orientation which would appear on side s of the tile when it has been reflected according to r. (E.g. for the tile type shown in Fig. 2, $S(H, W) = E$, and $S(H, N) = N$.) It is important to note that a glue does not have any particular orientation along the edge on which it resides, and so remains unchanged throughout reflections.

Two tiles t and t' that are placed at the points $\boldsymbol{a}$ and $\boldsymbol{a} + \boldsymbol{u}$ and reflected by $r \in R$ and $r' \in R$, respectively, *bind* with *strength* $\mathrm{str}_t\left(S(r, \boldsymbol{u})\right)$ if and only if $\left(\mathrm{label}_t\left(S(r, \boldsymbol{u})\right), \mathrm{str}_t\left(S(r, \boldsymbol{u})\right)\right) = \left(\overline{\mathrm{label}_{t'}\left(S(r', -\boldsymbol{u})\right)}, \mathrm{str}_{t'}\left(S(r', -\boldsymbol{u})\right)\right)$. That is, the glues on adjacent edges of two tiles bind iff they have complementary labels (usually specified by the same string with and without a trailing $'$) and the same strength.

In the following, given two partial functions f, g, we write $f(x) = g(x)$ if f and g are both defined and equal on x, or if f and g are both undefined on x.

Fix a finite set T of tile types. A *T-assembly*, sometimes denoted simply as an *assembly* when T is clear from the context, is a partial function $\alpha : \mathbb{Z}^2 \dashrightarrow T \times R$ defined on at least one input, with points $\boldsymbol{x} \in \mathbb{Z}^2$ at which $\alpha(\boldsymbol{x})$ is undefined interpreted to be empty space, so that dom α is the set of points with *oriented* tiles. We write $|\alpha|$ to denote $|\mathrm{dom}\ \alpha|$, and we say α is *finite* if $|\alpha|$ is finite. For a given location $\boldsymbol{v} \in \mathbb{Z}^2$, we denote the tile in α at location $\boldsymbol{v}$ by $\alpha(\boldsymbol{v})$ (if no tile exists there, $\alpha(\boldsymbol{v})$ is undefined). Let F be a function which takes as input an assembly α, a reflection $r \in R$, and a translation vector $\boldsymbol{v} \in \mathbb{Z}^2$, and which returns the assembly α^r, corresponding to α reflected according to r then translated by $\boldsymbol{v}$. We say that two assemblies, α and β, are equivalent iff there exists some reflection $r \in R$ and translation vector $\boldsymbol{v} \in \mathbb{Z}^2$ such that for $\beta' = F(\beta, r, \boldsymbol{v})$, $|\alpha| = |\beta'|$ and for all $\boldsymbol{v} \in |\alpha|$, $\alpha(\boldsymbol{v}) = \beta'(\boldsymbol{v})$. That is, α and β are equivalent iff one of them can be flipped and translated so that they perfectly match at all locations. For assemblies α and α', we say that α is a *subassembly* of α', and write $\alpha \sqsubseteq \alpha'$, if dom $\alpha \subseteq$ dom α' and $\alpha(\boldsymbol{x}) = \alpha'(\boldsymbol{x})$ for all $x \in$ dom α. An assembly α is *τ-stable* For some $\tau \in \mathbb{N}$, an assembly α is *τ-stable* if every cut of the binding graph of α has weight at least τ, where the weight of an edge is the strength of the glue it represents. When τ is clear from context, we say α is *stable*.

For a tile set T, we let $p_T : T \times R \to T$ be the projection map onto T (i.e. $p_T((t, r)) = t$). A *configuration* given by an assembly α is defined to be the map from $\mathbb{Z}^2$ to T given by $p_T \circ \alpha$.

Self-assembly begins with a *seed assembly* σ, in which each tile has a specified and fixed orientation, and proceeds asynchronously and nondeterministically, with tiles in any valid reflection in R adsorbing one at a time to the existing assembly in any manner that preserves τ-stability at all times. A *tile assembly system* (*TAS*) is an ordered triple $\mathcal{T} = (T, \sigma, \tau)$, where T is a finite set of tile types, σ is a seed assembly with finite domain in which each tile is given a fixed orientation, and $\tau \in \mathbb{N}$ is the *temperature*. A *generalized tile assembly system* (*GTAS*) is defined similarly, but without the finiteness requirements. We write $\mathcal{A}[\mathcal{T}]$ for the set of all assemblies that can arise (in finitely many steps or in the limit) from $\mathcal{T}$. An assembly $\alpha \in \mathcal{A}[\mathcal{T}]$ is *terminal*, and we write $\alpha \in \mathcal{A}_\square[\mathcal{T}]$, if no tile can be τ-stably added to it. It is clear that $\mathcal{A}_\square[\mathcal{T}] \subseteq \mathcal{A}[\mathcal{T}]$.

An assembly sequence in a TAS $\mathcal{T}$ is a (finite or infinite) sequence $\boldsymbol{\alpha} = (\alpha_0, \alpha_1, \dots)$ of assemblies in which each α_{i+1} is obtained from α_i by the addition of a single tile. The *result* $res(\boldsymbol{\alpha})$ of such an assembly sequence is its unique limiting assembly. (This is the last assembly in the sequence if the sequence is finite.) The set $\mathcal{A}[\mathcal{T}]$ is partially ordered by the relation $\longrightarrow$ defined by

$$\alpha \longrightarrow \alpha' \text{ iff there is an assembly sequence } \boldsymbol{\alpha} = (\alpha_0, \alpha_1, \dots)$$
$$\text{such that } \alpha_0 = \alpha \text{ and } \alpha' = res(\boldsymbol{\alpha}).$$

We say that $\mathcal{T}$ is *strongly directed* if and only if either $|\mathcal{A}_\square[\mathcal{T}]| = 1$ or if for every pair of terminal assemblies $\alpha, \beta \in \mathcal{A}_\square[\mathcal{T}]$, there exists a reflection $r \in R$ and a translation vector $\boldsymbol{v} \in \mathbb{Z}^2$ such that $\alpha = F(\beta, r, \boldsymbol{v})$. Furthermore, we say that $\mathcal{T}$ is *directed* if and only if for all $\alpha, \beta \in \mathcal{T}$, there exists a reflection $r \in R$ and a translation vector $\boldsymbol{v} \in \mathbb{Z}^2$ such that $p_T \circ \alpha = p_T \circ F(\beta, r, \boldsymbol{v})$. In other words, all of the assemblies of a directed systems give the same configuration.

A set $X \subseteq \mathbb{Z}^2$ *weakly self-assembles* if there exists a TAS $\mathcal{T} = (T, \sigma, \tau)$ and a set $B \subseteq T$ such that for each $\alpha \in \mathcal{A}_\square[\mathcal{T}]$ there exists a reflection $r \in R$ and a translation $\boldsymbol{v} \in \mathbb{Z}^2$ such that $\alpha_r = F(\alpha, r, \boldsymbol{v})$ and $\alpha_r^{-1}(B) = X$ holds. Essentially, weak self-assembly can be thought of as the creation (or "painting") of a pattern of tiles from B (usually taken to be a unique "color" such as black) on a possibly larger "canvas" of un-colored tiles.

A set X *strictly self-assembles* if there is a TAS $\mathcal{T}$ such that for each assembly $\alpha \in \mathcal{A}_\square[\mathcal{T}]$ there exists a reflection $r \in R$ and a translation $\boldsymbol{v} \in \mathbb{Z}^2$ such that $\alpha_r = F(\alpha, r, \boldsymbol{v})$ and dom $\alpha_r = X$. Essentially, strict self-assembly means that tiles are only placed in positions defined by X. Note that if X strictly self-assembles, then X weakly self-assembles. X in the definition of strict or weak self-assembly is called a *shape* in $\mathbb{Z}^2$.

In this paper, we also consider scaled-up versions of shapes. Formally, if X is a shape and $c \in \mathbb{N}$, then a *c-scaling* of X is defined as the set $X^c = \{(x, y) \in \mathbb{Z}^2 \mid (\lfloor \frac{x}{c} \rfloor, \lfloor \frac{y}{c} \rfloor) \in X\}$. Intuitively, X^c is the shape obtained by replacing each point in X with a $c \times c$ block of points. We refer to the natural number c as the *scale factor*.

2.1   Paths in the Binding Graph and as Assemblies

Given an assembly α and locations $\boldsymbol{x}$ and $\boldsymbol{y}$ such that $\boldsymbol{x}, \boldsymbol{y} \in \text{dom } \alpha$, we define a *path in* α *from* $\boldsymbol{x}$ *to* $\boldsymbol{y}$ (or simply a *path from* $\boldsymbol{x}$ *to* $\boldsymbol{y}$) as a simple directed path in the binding graph of α with the first location being $\boldsymbol{x}$ and the last $\boldsymbol{y}$. We refer to such a path as π_x^y, and for $k = |\pi_x^y|$ (i.e. k is the length of, or number of tiles on, π_x^y) and $0 \leq i < k$, let $\pi_x^y(i)$ be the ith location of π_x^y. Thus, $\pi_x^y(0) = \boldsymbol{x}$, and $\pi_x^y(k-1) = \boldsymbol{y}$. We can thus refer to the ith tile on π_x^y and its reflection as $\alpha(\pi_x^y(i))$, and as shorthand will often refer to locations and/or tiles along a path. Regardless of the order in which the tiles of π_x^y were placed in α, we define *input* and *output* sides for each tile in π_x^y (except for the first and last, respectively) in relation to their position on π_x^y. The input side of the ith tile of π_x^y, $\alpha(\pi_x^y(i))$, is that which binds to $\alpha(\pi_x^y(i-1))$, and the output side is that which binds to $\alpha(\pi_x^y(i+1))$. We denote these sides as $IN(\pi_x^y(i))$ and $OUT(\pi_x^y(i))$, respectively. (Thus, $\alpha(\pi_x^y(0))$ has no input side, and $\alpha(\pi_x^y(k-1))$ has no output side.) Note that in a temperature-1 system, an assembly α' exactly representing π_x^y would be able to grow solely from $\alpha(\pi_x^y(0))$, in the order of π_x^y, with each tile having input and output sides as defined for π_x^y.

3   The RTAM is Not Computationally Universal at $\tau = 1$

In this section, we show that directed RTAM systems are not computationally universal by showing that any shape weakly assembled by a directed RTAM system is "simple". We will first define our notion of simple. Many of the following definitions can also be found in [7].

Definition 1. *A set* $X \subseteq \mathbb{Z}^2$ *is* semi-doubly periodic *if there exist three vectors* $\boldsymbol{b}$, $\boldsymbol{u}$, *and* $\boldsymbol{v}$ *in* $\mathbb{Z}^2$ *such that*

$$X = \{\ \boldsymbol{b} + n \cdot \boldsymbol{u} + m \cdot \boldsymbol{v} \mid n, m \in \mathbb{N}\ \}.$$

Less formally, a semi-doubly periodic (a.k.a linear) set is a set that repeats infinitely along two vectors (linearly independent vectors in the non-degenerate case), starting at some base point $\boldsymbol{b}$. Now, let $\mathcal{T} = (T, \sigma, 1)$ refer to a directed, temperature-1 RTAM system. We show that any such $\mathcal{T}$ weakly self-assembles a set $X \subseteq \mathbb{Z}^2$ that is a finite union of semi-doubly periodic sets (a.k.a. semilinear).

Theorem 1. *Let* $\mathcal{T} = (T, \sigma, 1)$ *be a directed RTAM system. If a set* $X \subseteq \mathbb{Z}^2$ *weakly self-assembles in* $\mathcal{T}$, *then* X *is a finite union of semi-doubly periodic sets.*

Proof. (sketch) Here we give a high-level sketch of the proof of Theorem 1. See [11] for a rigorous proof. The basic idea of the proof is as follows. For an RTAM system $\mathcal{T} = (T, \sigma, 1)$ we consider all of the paths of n tiles (for n to be defined) that can assemble from each exposed glue of σ such that each consecutive tile that binds forming the path attaches via a north or west glue (and we say that such a path "extends to the north-west") (Fig. 3).

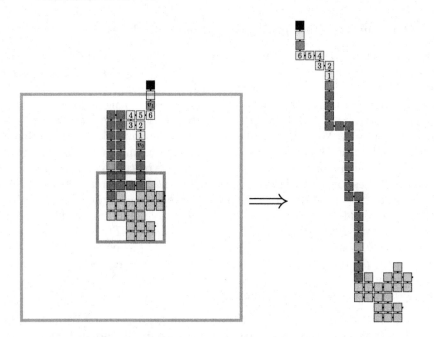

Fig. 3. A depiction of tiles for a path that can assemble in $\mathcal{T}$. The original path is on the left. The path on the right is a modification to the path on the left that must also be able to assemble in $\mathcal{T}$. The blue tiles labeled v_0 and v_1 are of the same tile type and orientation. The tiles labeled 1 through 6 can be repeated indefinitely as depicted in Table 1(a).

Any finite path is trivially the union of semi-doubly periodic sets. Then, for n sufficiently large, a path of n tiles that extends to the north-west must contain two distinct tiles t_1 and t_2 of the same tile type in the same orientation. If for every such path, every two distinct tiles t_1 and t_2 of the same tile type in the same orientation lie on a horizontal or vertical line, then it can be argued that the terminal configuration of $\mathcal{T}$ must consist of finitely many infinitely long horizontal or vertical paths connected to σ, and is therefore the finite union of semi-doubly periodic sets. On the other hand, if there is a path such that the two distinct tiles t_1 and t_2 of the same tile type in the same orientation do not lie on a horizontal or vertical path, then we argue that the terminal assembly of $\mathcal{T}$ is the finite union of semi-doubly periodic sets as follows. First, we note that for such a path, π say, the tiles between t_1 and t_2 can be repeated indefinitely. This is shown in Table 1(a). Then we show how to modify π by reflecting tiles to obtain an infinite family of paths. Examples of such modified paths are shown in Table 1(b)–(h). Now, if a tile t belongs to one of these paths and has location l say, then, since $\mathcal{T}$ is directed the terminal assembly of $\mathcal{T}$ must contain a tile of the same type as t at each such location l. Finally, we note that all of these paths taken together form a semi-doubly periodic set. This is depicted in Fig. 4a. Continuing this line of reasoning, we show that the terminal assembly of $\mathcal{T}$ is the finite union of semi-doubly periodic sets. A portion of such an assembly is shown in Fig. 4b.   □

Table 1. Each figure in this table depicts a possible path that can assemble in $\mathcal{T}$. The yellow tiles make up the seed and the green tiles are a path leading to the repeated tile that allows the path to repeat.

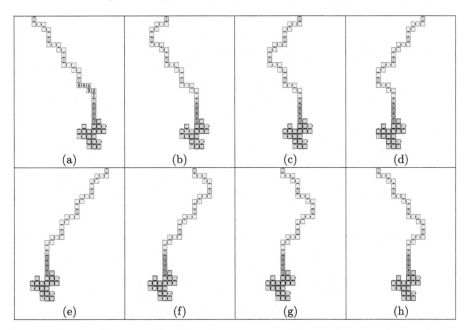

An intuitive reason that Theorem 1 supports the conclusion that the RTAM is not computationally universal is as follows. Let $H = \{(i, x) \mid$ program i halts when run on input $x\}$ be the halting set and let M be a Turing machine that outputs a 1 if $(i, x) \in H$. The typical way of expressing the computation of M in tile assembly is as follows. For a fixed tileset T, a seed assembly σ encodes an "input" to the computation, while the "output" of 1 by M corresponds to the translation of some configuration being contained in the terminal assembly α_σ of $(T, \sigma, 1)$. For this sense of computation, the following corollary says that the set of seed assemblies that "output" a 1 is a recursive set (not just a recursively enumerable set). This would contradict the fact that the halting set is not recursive. This is stated in the following corollary. For a more formal statement of this corollary and a proof, see [11].

Corollary 1. *For any tileset T in the RTAM and fixed finite configuration C, let S be the set of seed assemblies σ such that (1) the RTAM system $(T, \sigma, 1)$ is directed and (2) the terminal assembly of T contains C. Then, S is a recursive set.*

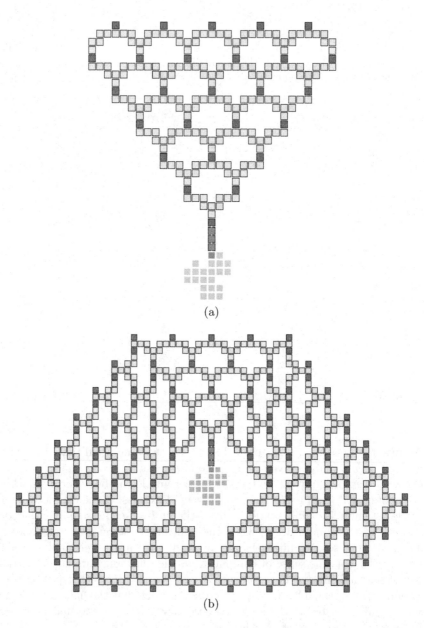

Fig. 4. (a) A configuration that can be thought of as the "union" of the type-consistent assemblies depicted in Table 1. (b) A configuration of tiles that must weakly self-assemble in $\mathcal{T}$.

3.1   Universal Computation at $\tau = 2$

In this section give a theorem that states that universal computation is possible in the RTAM at temperature 2. We give an example of simulating a binary counter in the RTAM and give the general proof in [11].

Theorem 2. *The RTAM is computationally universal at $\tau = 2$. Moreover, the class of directed RTAM systems is computationally universal at $\tau = 2$.*

First we note that given a Turing machine M, we use Lemma 7 of [2] to obtain a tile set which simulates M using a *zig-zag system*. In fact, as noted in [17], we can find a singly seeded *compact zig-zag system* $\mathcal{T} = (T, \sigma, 2)$ with $\mathcal{A}_\square[\mathcal{T}] = \{\alpha\}$ which simulates M. Then the proof of Theorem 2 relies on showing that any compact zig-zag system in the aTAM at temperature 2 can be converted into a directed RTAM system $\mathcal{S}$ that is "almost" compact zig-zag. The RTAM system that we construct differs from a compact zig-zag system in that when the length of a row of the growing zig-zag assembly increases by a tile, a strength-2 glue is exposed that allows a tile to bind below the row. This results in the possibility of a single "misplaced" tile per row, but nevertheless, this is enough to simulate a Turing machine. The proof of Theorem 2 can be found in [11].

4   Self-assembly of Shapes in the RTAM

In this section, we discuss the self-assembly of shapes in the RTAM, especially the commonly used benchmark of squares. At temperature 2, using a zig-zag binary counter similar to that used in Sect. 3.1, $n \times n$ squares can be built using the optimal $\log n/(\log \log n)$ tile types following the construction of [21] with only trivial modifications. Similarly, the majority of shapes which can be weakly self-assembled in the temperature-2 aTAM can be built in the temperature-2 RTAM, although shapes with single-tile-wide branches which are not symmetric are impossible to strictly self-assemble in the RTAM.

At temperature-1, however, the differences between the powers of the aTAM and RTAM appear to increase. Here we will demonstrate that squares whose sides are of even length cannot weakly (or therefore strictly) self-assemble in the RTAM at $\tau = 1$, although any square can strictly self-assemble in the $\tau = 1$ aTAM. We then prove a tight bound of n tile types required to self-assemble an $n \times n$ square for odd n in the RTAM at $\tau = 1$. (Which is, interestingly, better than the conjectured lower bound of $2n - 1$ for the $\tau = 1$ aTAM).

4.1   For Even n, No $n \times n$ Square Self-assembles in the $\tau = 1$ RTAM

Theorem 3. *For all $n \in \mathbb{Z}^+$ where n is even, there exists no RTAM system $\mathcal{T} = (T, \sigma, 1)$ where $|\sigma| = 1$ and $\mathcal{T}$ weakly (or strictly) self-assembles an $n \times n$ square.*

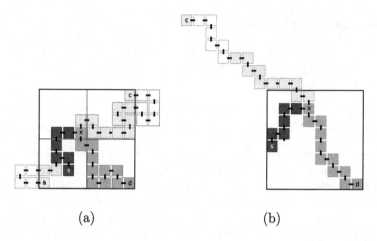

Fig. 5. (a) Example paths in an 8×8 square (i.e. of even dimension). The path from corners *a* and *c* is composed of tiles of different colors. The path from corner *d* to a point on that path is dark grey. The path from the seed to that intersection is in black. The path to be stretched out is outlined in red. (b) The stretched out version of the paths (Color figure online).

Proof. (sketch) We prove Theorem 3 by contradiction, and here give a sketch of the proof. (See [11] for the full proof.) Therefore, assume that for some $n \in \mathbb{Z}^+$ such that $(n \mod 2) = 0$, there exists an RTAM system $\mathcal{T} = (T, \sigma, 1)$ such that $|\sigma| = 1$ and $\mathcal{T}$ weakly self-assembles an $n \times n$ square S. We take $\alpha \in \mathcal{A}_\square[\mathcal{T}]$ and consider the corners of the square which is weakly self-assembled.

There must exist a path which connects two diagonal corners (*a* to *c* in Fig. 5 (a)), and that path must travel through 3 quadrants of the square since the dimensions are even length and diagonal paths are not possible in the grid graph of an assembly. We then find a path connecting the corner of the quadrant (possibly) unvisited by that path (e.g. *d*), and note that since α must be connected that we can find a new path which connects the new corner (*d*) to one of the original corners (*a* or *c*) via a path which crosses across the midpoint of the square from either corner on the new path. Finally, we demonstrate that there is an assembly sequence which starts from the seed and grows to that new path, then builds that path in a way such that it is maximally "stretched" out by appropriately flipping the tiles. (See Fig. 5(b) for an example.) This stretched out path is producible by a valid assembly sequence but must grow beyond the bounds of the square (by at least one position), so $\mathcal{T}$ does not weakly (or strictly) self-assemble the square.

4.2   Tight Bounds on the Tile Complexity of Squares of Odd Dimension

In this section, we prove tight bounds on the number of tiles necessary to self-assemble a square of odd dimension.

Theorem 4. *For all $n \in \mathbb{Z}^+$ where n is odd, an $n \times n$ square strictly self-assembles in an RTAM system $\mathcal{T} = (T, \sigma, 1)$ where $|T| = n$ and $|\sigma| = 1$.*

We prove Theorem 4 by giving a scheme for obtaining the tileset for any given n that exploits the fact that for n odd, an $n \times n$ square in $\mathbb{Z}^2$ is symmetric across a row and column points. Although Theorem 4 pertains to squares, a simple modification of the proof shows the following corollary.

Corollary 2. *For all $n, m \in \mathbb{Z}^+$ where n is odd, an $n \times m$ square strictly self-assembles in an RTAM system $\mathcal{T} = (T, \sigma, 1)$ where $|T| = \frac{n+m}{2}$ and $|\sigma| = 1$.*

We also prove that the upper bound of Theorem 4 is tight, i.e. an $n \times n$ square, where n is odd, cannot be self-assembled using less than n tile types. The proof of this theorem can be found in [11].

Theorem 5. *For all $n \in \mathbb{Z}^+$ where n is odd, there exists no RTAM system $\mathcal{T} = (T, \sigma, 1)$ where $|T| < n$ and $|\sigma| = 1$ such that $\mathcal{T}$ weakly (or strictly) self-assembles an $n \times n$ square.*

4.3 Assembling Finite Shapes in the RTAM

In this section we first give a corollary of Theorem 4 showing that sufficiently symmetric shapes weakly self-assemble in the RTAM. Then we prove 3 theorems about assembling finite shapes in $\mathbb{Z}^2$ in the RTAM. These theorems show that the assembly of finite shapes in singly seeded RTAM systems is quite a bit different than the assembly of finite shapes in the aTAM by singly seeded systems.

Given a shape S in $\mathbb{Z}^2$, let χ_S denote the characteristic function of the set S. That is, $\chi_S(x, y) = 1$ if $x \in S$ and $\chi_S(x, y) = 0$ otherwise. Then, we say that a shape S is *odd-symmetric* with respect to a horizontal line $y = l$ (respectively, vertical line $x = l$) for $l \in \mathbb{Z}^2$ iff for all $(a, b) \in \mathbb{Z}^2$, $\chi_S(a, l - b) = \chi_S(a, l + b)$ (respectively, $\chi_S(l - a, b) = \chi_S(l + a, b)$). If there exists a line such that a shape, S, is odd-symmetric with respect to this line, we say that the shape is odd-symmetric. Given a shape S, we call the smallest rectangle of points in $\mathbb{Z}^2$ containing S the *bounding-box* for S.

Let R denote the bounding box of an odd-symmetric shape S, and let n and m in $\mathbb{N}$ be the dimensions of R. A simple modification to the proof of Theorem 4 where a tile is labeled with a label B if and only if it corresponds to points in S, shows that odd-symmetric shapes can weakly assemble in the RTAM.

Corollary 3. *Given a shape S in $\mathbb{Z}^2$, if S is odd-symmetric, then there exists an RTAS $\mathcal{T} = (T, \sigma, \tau)$ such that $|\sigma| = 1$, $\tau \geq 1$, and $\mathcal{T}$ weakly assemblies S.*

Additionally, if one is willing to build 2 mirrored copies of the shape in each assembly, then any finite shape can be weakly self-assembled in the RTAM at $\tau = 1$, along with its mirrored copy (at a cost of tile complexity approximately equal to the number of points in the shape) by simply building a central column (or row) from which identical copies of hardcoded rows (or columns) grow, so that each side grows a reflected copy of the shape in hardcoded slices.

We say that a TAS (in either the aTAM or the RTAM) $\mathcal{T}$ is called *mismatch-free* if for every producible assembly $\alpha \in \mathcal{A}[\mathcal{T}]$ with two neighboring tiles with abutting edges e_1 and e_2, either e_1 and e_2 do not have glues or e_1 and e_2 have glues with matching labels and strengths. Then, for singly seeded aTAM systems, any finite connected shape can be strictly assembled by a mismatch-free system. Theorems 6, 7, and 8 show that assembling shapes in the RTAM is more complex. The proofs of these theorems can be found in [11].

Theorem 6. *There exists a finite connected shape S in $\mathbb{Z}^2$ that weakly self-assembles in a singly seeded RTAM system such that there exists no singly seeded RTAM system that strictly self-assembles S.*

Theorem 7. *There exists a finite shape S in $\mathbb{Z}^2$ that can be strictly self-assembled by some singly seeded RTAM system such that every singly seeded RTAM system at temperature 1 which strictly self-assembles S is not directed.*

Theorem 8. *There exists a finite shape S in $\mathbb{Z}^2$ such that every singly seeded RTAM system that strictly self-assembles S is not mismatch-free.*

4.4   Mismatch-Free Assembly of Finite Shapes in the RTAM

Given a shape S, i.e. a finite connected subset of $\mathbb{Z}^2$, we say that a *graph of S* is a graph $G_S = (V, E)$ with a vertex at the center of each point in S and an edge between every pair of vertices at adjacent points of S. A *tree of S*, T_S, is a graph of S which is a tree. (See Fig. 6 for examples of S, G_S, and G_T.) Given a graph $G = (V, E)$, we say that an *axis* of G is a horizontal or vertical line of vertices such that there is an edge between each pair of adjacent points on that line. Notice that two distinct axes can be collinear. Given an axis a, an *axial branch* of T_S is a branch of T_S containing exactly one vertex v on a and all vertices and edges of T_S which are connected to a vertex that does not lie on a and is adjacent to v. We say that the branch *begins from v*. Intuitively, an axial branch is a connected component extending from an axis. (See the pink highlighted portion of Fig. 6c for an example axial branch off the green axis).

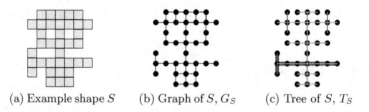

(a) Example shape S (b) Graph of S, G_S (c) Tree of S, T_S

Fig. 6. An example ϵ-symmetric shape (Color figure online).

A tree T_S is symmetric across an axis a if, for every vertex v contained on a, the branches of a which begin from v are symmetric across a. A tree T_S is *off-by-one symmetric* across an axis a if, for every vertex v except for at most 1, the branches of a which begin from v are symmetric across a. See Fig. 6c for an example of such a tree, with the axis a shown in green.

Definition 2. *A tree T is ϵ-symmetric if and only if for any axis a of T, T is off-by-one symmetric across a.*

Definition 3. *Given a shape S with graph G_S, we say that S is ϵ-symmetric if and only if there exists a spanning tree, T_S, of G_S such that T_S is ϵ-symmetric.*

For an example of an ϵ-symmetric shape S, see Fig. 6a. The tree T_S is off-by-one symmetric across the vertical green axis, the branches off of that axis are symmetric across the horizontal yellow axes, and the branches off of those axes are symmetric across the vertical blue axes. The following theorem gives a complete classification of finite connected shapes which can be assembled by temperature-1 singly seeded mismatch-free RTAM systems. The proof of this theorem is given in [11].

Theorem 9. *Let $S \subset \mathbb{Z}^2$ be a finite connected shape. There exists a mismatch-free RTAM system $T = (T, \sigma, 1)$ with $|\sigma| = 1$ that strictly assembles S if and only if S is ϵ-symmetric.*

While Theorem 9 shows exactly which shapes can be assembled without cooperation or mismatches by singly seeded RTAM systems, the following theorem shows that with cooperation, RTAM systems can assemble arbitrary scale factor 2 shapes. The proof of this theorem is in [11].

Theorem 10. *Let $S \subset \mathbb{Z}^2$ be a finite connected shape, and S^2 be S at scale factor 2. There exists a mismatch-free RTAM system $T = (T, \sigma, 2)$ with $|\sigma| = 1$ that strictly self-assembles S^2.*

References

1. Barish, R.D., Schulman, R., Rothemund, P.W.K., Winfree, E.: An information-bearing seed for nucleating algorithmic self-assembly. Proc. Natl. Acad. Sci. **106**(15), 6054–6059 (2009)
2. Cook, M., Fu, Y., Schweller, R.T.: Temperature 1 self-assembly: deterministic assembly in 3D and probabilistic assembly in 2D. In: SODA 2011: Proceedings of the 22nd Annual ACM-SIAM Symposium on Discrete Algorithms. SIAM (2011)
3. Santini, C.C., Bath, J., Tyrrell, A.M., Turberfield, A.J.: A clocked finite state machine built from DNA. Chem. Commun. **49**, 237–239 (2013)
4. Demaine, E.D., Demaine, M.L., Fekete, S.P., Patitz, M.J., Schweller, R.T., Winslow, A., Woods, D.: One tile to rule them all: simulating any tile assembly system with a single universal tile. In: Esparza, J., Fraigniaud, P., Husfeldt, T., Koutsoupias, E. (eds.) ICALP 2014. LNCS, vol. 8572, pp. 368–379. Springer, Heidelberg (2014)
5. Doty, D.: Theory of algorithmic self-assembly. Commun. ACM **55**(12), 78–88 (2012)

6. Doty, D., Kari, L., Masson, B.: Negative interactions in irreversible self-assembly. Algorithmica **66**(1), 153–172 (2013)
7. Doty, D., Patitz, M.J., Summers, S.M.: Limitations of self-assembly at temperature 1. Theoret. Comput. Sci. **412**, 145–158 (2011)
8. Fekete, S.P., Hendricks, J., Patitz, M.J., Rogers, T.A., Schweller, R.T.: Universal computation with arbitrary polyomino tiles in non-cooperative self-assembly. In: Proceedings of the Twenty-Sixth Annual ACM-SIAM Symposium on Discrete Algorithms (SODA 2015), San Diego, CA, USA 4–6 January 2015, pp. 148–167 (2015)
9. Fu, B., Patitz, M.J., Schweller, R.T., Sheline, R.: Self-assembly with geometric tiles. In: Czumaj, A., Mehlhorn, K., Pitts, A., Wattenhofer, R. (eds.) ICALP 2012, Part I. LNCS, vol. 7391, pp. 714–725. Springer, Heidelberg (2012)
10. Han, D., Pal, S., Yang, Y., Jiang, S., Nangreave, J., Liu, Y., Yan, H.: DNA gridiron nanostructures based on four-arm junctions. Science **339**(6126), 1412–1415 (2013)
11. Hendricks, J., Patitz, M.J., Rogers, T.: Reflections on tiles (in self-assembly). Technical Report 1404.5985, Computing Research Repository (2014)
12. Ke, Y., Ong, L.L., Shih, W.M., Yin, P.: Three-dimensional structures self-assembled from DNA bricks. Science **338**(6111), 1177–1183 (2012)
13. Kim, J.-W., Kim, J.-H., Deaton, R.: DNA-linked nanoparticle building blocks for programmable matter. Angew. Chem. Int. Ed. **50**(39), 9185–9190 (2011)
14. Lathrop, J.I., Lutz, J.H., Summers, S.M.: Strict self-assembly of discrete Sierpinski triangles. Theoret. Comput. Sci. **410**, 384–405 (2009)
15. Mao, C., LaBean, T.H., Relf, J.H., Seeman, N.C.: Logical computation using algorithmic self-assembly of DNA triple-crossover molecules. Nature **407**(6803), 493–6 (2000)
16. Patitz, M.J.: An introduction to tile-based self-assembly and a survey of recent results. Nat. Comput. **13**(2), 195–224 (2014)
17. Patitz, M.J., Schweller, R.T., Summers, S.M.: Exact shapes and turing universality at temperature 1 with a single negative glue. In: Cardelli, L., Shih, W. (eds.) DNA 17 2011. LNCS, vol. 6937, pp. 175–189. Springer, Heidelberg (2011)
18. Pinheiro, A.V., Han, D., Shih, W.M., Yan, H.: Challenges and opportunities for structural DNA nanotechnology. Nat. Nanotechnol. **6**(12), 763–772 (2011)
19. Rothemund, P.W.K.: Theory and experiments in algorithmic self-assembly. Ph.D. thesis, University of Southern California, December 2001
20. Rothemund, P.W.K., Papadakis, N.: E., Winfree: Algorithmic self-assembly of dna sierpinski triangles. PLoS Biol. **2**(12), e424 (2004)
21. Rothemund, P.W.K., Winfree, E.: The program-size complexity of self-assembled squares (extended abstract). In: STOC 2000: Proceedings of the Thirty-second Annual ACM Symposium on Theory of Computing, Portland, Oregon, USA, pp. 459–468. ACM (2000)
22. Schulman, R., Winfree, E.: Synthesis of crystals with a programmable kinetic barrier to nucleation. Proc. Natl. Acad. Sci. **104**(39), 15236–15241 (2007)
23. Winfree, E.: Algorithmic self-assembly of DNA. Ph.D. thesis, California Institute of Technology, June 1998
24. Winfree, E., Liu, F., Wenzler, L.A., Seeman, N.C.: Design and self-assembly of two-dimensional DNA crystals. Nature **394**(6693), 539–44 (1998)

Optimal Program-Size Complexity
for Self-Assembly at Temperature 1 in 3D

David Furcy[1], Samuel Micka[2], and Scott M. Summers[3]([✉])

[1] Department of Computer Science, University of Wisconsin-Oshkosh,
Oshkosh, WI 54901, USA
`furcyd@uwosh.edu`
[2] Computer Science Department, Montana State University,
Bozeman, MT 59717, USA
`sam.micka@cs.montana.edu`
[3] Department of Computer Science, University of Wisconsin-Oshkosh,
Oshkosh, WI 54901, USA
`summerss@uwosh.edu`

Abstract. Working in a three-dimensional variant of Winfree's abstract Tile Assembly Model, we show that, for all $N \in \mathbb{N}$, there is a tile set that uniquely self-assembles into an $N \times N$ square shape at temperature 1 with optimal program-size complexity of $O(\log N / \log \log N)$ (the program-size complexity, also known as tile complexity, of a shape is the minimum number of unique tile types required to uniquely self-assemble it). Moreover, our construction is "just barely" 3D in the sense that it works even when the placement of tiles is restricted to the $z = 0$ and $z = 1$ planes. This result affirmatively answers an open question from Cook, Fu, Schweller (SODA 2011). To achieve this result, we develop a general 3D temperature 1 optimal encoding construction, reminiscent of the 2D temperature 2 optimal encoding construction of Soloveichik and Winfree (SICOMP 2007), and perhaps of independent interest.

1 Introduction

The simplest mathematical model of nanoscale tile self-assembly is Erik Winfree's abstract Tile Assembly Model (aTAM) [10]. The aTAM extends classical Wang tiling [9] in that the former bestows upon the latter a mechanism for sequential "growth" of a tile assembly. Very briefly, in the aTAM, the fundamental components are un-rotatable, translatable square "tile types" whose sides are labeled with (alpha-numeric) glue "colors" and (integer) "strengths". Two tiles that are placed next to each other *bind* if both the glue colors and the strengths on their abutting sides match and the sum of their matching strengths sum to at least a certain (integer) "temperature". Self-assembly starts from a "seed" tile type, typically assumed to be placed at the origin, and proceeds nondeterministically and asynchronously as tiles bind to the seed-containing assembly one at a time. In this paper, we work in a three-dimensional variant of the aTAM in which tile types are unit cubes and growth proceeds in a *noncooperative* manner.

© Springer International Publishing Switzerland 2015
A. Phillips and P. Yin (Eds.): DNA 2015, LNCS 9211, pp. 71–86, 2015.
DOI: 10.1007/978-3-319-21999-8_5

Tile self-assembly in which tiles may be placed in a noncooperative fashion is often referred to as "temperature 1 self-assembly". Despite the arcane name, this is a fundamental and ubiquitous form of growth: it refers to growth from *growing and branching tips* in Euclidean space, where each new tile is added if it can bind on at least *one side*. Note that a more general form of *cooperative* growth, where some of the tiles may be required to bind on two or more sides, leads to highly non-trivial behavior in the aTAM, e.g., Turing universality [10] and the efficient self-assembly of $N \times N$ squares [1,7] and other algorithmically specified shapes [8]. Doty, Patitz and Summers conjecture [3] that the shape or pattern produced by any 2D temperature 1 tile set that uniquely produces a final structure is "simple" in the sense of Presburger arithmetic [6]. However, their conjecture is currently unproven and it remains to be seen if noncooperative self-assembly in the aTAM can achieve the same computational and geometric expressiveness as that of cooperative self-assembly. In this paper, we focus on a (barely) 3D version of the problem of finding the minimum number of distinct tile types required to self-assemble an $N \times N$ square, i.e., its *tile complexity* (or *program-size complexity*), at temperature 1.

The tile complexity of an $N \times N$ square at temperature 1 has been studied extensively. In 2000, Rothemund and Winfree [7] proved that the tile complexity of an $N \times N$ square at temperature 1 is N^2, assuming the final structure is fully connected, and at most $2N - 1$, otherwise (they also conjectured that the lower bound, in general, is $2N - 1$). A decade later, Manuch, Stacho and Stoll [5] established that, assuming no strength/label mismatches between adjacent glues are present in the final assembly, the tile complexity of an $N \times N$ square at temperature 1 is $2N - 1$. Shortly thereafter, and quite surprisingly, Cook, Fu and Schweller [2] showed that the tile complexity of an $N \times N$ square at temperature 1 is $O(\log N)$ if tiles are allowed to be placed in the $z = 0$ and $z = 1$ planes (here, an $N \times N$ square is actually a full 2D square in the $z = 0$ plane with additional tiles above it in the $z = 1$ plane).

Technically speaking, the aforementioned, just-barely-3D construction of Cook, Fu and Schweller is actually a general transformation that takes as input a 2D temperature 2 "zig-zag" tile set, say T, and outputs a corresponding 3D temperature 1 tile set, say T', that simulates T. In this transformation from T to T', the tile complexity increases by $O(\log g)$, where g is the number of unique north/south glues in the input tile set T. Since the number of north/south glues in the standard 2D aTAM base-2 binary counter is $O(1)$, Cook, Fu and Schweller use their transformation to produce several tile sets, which, when wired together appropriately and combined with "filler" tiles, self-assemble into an $N \times N$ square at temperature 1 in 3D with $O(\log N)$ tile complexity.

Of course, it is well-known that the tile complexity of an $N \times N$ square at temperature 2 is $O\left(\frac{\log N}{\log \log N}\right)$ [1], which, as Cook, Fu and Schweller point out in [2], is achievable using a zig-zag counter with an optimally-chosen base, say b, which satisfies $\frac{\log N}{\log \log N} \leq b < \frac{2 \log N}{\log \log N}$, rather than in base $b = 2$. However, using currently-known techniques, counting in base b at temperature 2 requires having a tile set with $\Theta(b)$ unique north/south glues, whence the zig-zag transformation of Cook,

Fu and Schweller cannot be used to get $O\left(\frac{\log N}{\log \log N}\right)$ tile complexity for an $N \times N$ square at temperature 1 in 3D. Moreover, the *optimal encoding* scheme of Soloveichik and Winfree [8] and the *base conversion* technique of Adleman et al. [1] do not work correctly at temperature 1 and they also cannot be simulated by the Cook, Fu and Schweller construction without an $\Omega\left(\frac{\log N}{\log \log N}\right)$ blowup in tile complexity. Thus, Cook, Fu and Schweller, at the end of Sect. 4.4 in [2], pose the following question: is it possible to achieve the tile complexity bound of $O\left(\frac{\log N}{\log \log N}\right)$ for an $N \times N$ square at temperature 1 in 3D?

In Theorem 1, the main theorem of this paper, we answer the previous question in the affirmative, i.e., we prove that the tile complexity of an $N \times N$ square at temperature 1 in 3D is $O\left(\frac{\log N}{\log \log N}\right)$ (in our construction, tiles are placed only in the $z = 0$ and $z = 1$ planes of $\mathbb{Z}^3$). Our tile complexity matches a corresponding lower bound dictated by Kolmogorov complexity (see [4] for details on Kolmogorov complexity), which was established by Rothemund and Winfree in 2000, and holds for all "algorithmically random" values of N [7][1]. Thus, our construction yields optimal tile complexity for the self-assembly of $N \times N$ squares at temperature 1 in 3D, for all algorithmically random values of N. To achieve optimal tile complexity, we adapt the optimal encoding technique of Soloveichik and Winfree [8] (which, itself, is based on the base-conversion scheme of [1]) to work at temperature 1 in 3D. Our 3D temperature 1 optimal encoding technique, described in Sect. 3, is perhaps of independent interest.

2   Definitions: 3D Abstract Tile Assembly Model

In this section, we give a brief sketch of a 3-dimensional version of Winfree's abstract Tile Assembly Model.

Let Σ be an alphabet. A 3-dimensional *tile type* is a tuple $t \in (\Sigma^* \times \mathbb{N})^6$, i.e., a unit cube with six sides listed in some standardized order, each side having a *glue* $g \in \Sigma^* \times \mathbb{N}$ consisting of a finite string *label* and a non-negative integer *strength*. In this paper, all glues have strength 1. There is a finite set T of 3-dimensional tile types but an infinite number of copies of each tile type, with each copy being referred to as a *tile*.

A 3-dimensional *assembly* is a positioning of tiles on the integer lattice $\mathbb{Z}^3$ and is described formally as a partial function $\alpha : \mathbb{Z}^3 \dashrightarrow T$. Two adjacent tiles in an assembly *bind* if the glue labels on their abutting sides are equal and have positive strength. Each assembly induces a *binding graph*, i.e., a (square) grid graph whose vertices are positions of tiles and whose edges connect any two vertices whose corresponding tiles bind. If τ is an integer, we say that an assembly is τ-*stable* if every cut of its binding graph has strength at least τ, where the strength of a cut is the sum of all of the individual glue strengths in the cut.

---

[1] Technically, Rothemund and Winfree established the 2D self-assembly case, but their proof easily generalizes to 3D self-assembly.

A 3-dimensional *tile assembly system* (TAS) is a triple $\mathcal{T} = (T, \sigma, \tau)$, where T is a finite set of tile types, $\sigma : \mathbb{Z}^3 \dashrightarrow T$ is a finite, τ-stable *seed assembly*, and τ is the *temperature*. In this paper, we assume that $|\text{dom } \sigma| = 1$ and $\tau = 1$. An assembly α is *producible* if either $\alpha = \sigma$ or if β is a producible assembly and α can be obtained from β by the stable binding of a single tile. In this case we write $\beta \rightarrow_1^{\mathcal{T}} \alpha$ (to mean α is producible from β by the binding of one tile), and we write $\beta \rightarrow^{\mathcal{T}} \alpha$ if $\beta \rightarrow_1^{\mathcal{T}*} \alpha$ (to mean α is producible from β by the binding of zero or more tiles). When $\mathcal{T}$ is clear from context, we may write $\rightarrow_1$ and $\rightarrow$ instead. We let $\mathcal{A}[\mathcal{T}]$ denote the set of producible assemblies of $\mathcal{T}$. An assembly is *terminal* if no tile can be τ-stably bound to it. We let $\mathcal{A}_\square[\mathcal{T}] \subseteq \mathcal{A}[\mathcal{T}]$ denote the set of producible, terminal assemblies of $\mathcal{T}$.

A TAS $\mathcal{T}$ is *directed* if $|\mathcal{A}_\square[\mathcal{T}]| = 1$. Hence, although a directed system may be nondeterministic in terms of the order of tile placements, it is deterministic in the sense that exactly one terminal assembly is producible. For a set $X \subseteq \mathbb{Z}^3$, we say that X is *uniquely produced* if there is a directed TAS $\mathcal{T}$, with $\mathcal{A}_\square[\mathcal{T}] = \{\alpha\}$, and dom $\alpha = X$.

For $N \in \mathbb{N}$, we say that $S_N^3 \subseteq \mathbb{Z}^3$ is a 3D $N \times N$ *square* if $\{0, \ldots, N-1\} \times \{0, \ldots, N-1\} \times \{0\} \subseteq S_N^3 \subseteq \{0, \ldots, N-1\} \times \{0, \ldots, N-1\} \times \{0,1\}$. In other words, a 3D $N \times N$ square is at most two 2D $N \times N$ squares, one stacked on top of the other.

In the spirit of [7], we define the *tile complexity* of a 3D $N \times N$ square at temperature τ, denoted by $K_{3DSA}^\tau(N)$, as the minimum number of distinct 3D tile types required to uniquely produce it, i.e., $K_{3DSA}^\tau(N) = \min\{n \mid \mathcal{T} = (T, \sigma, \tau), |T| = n \text{ and } \mathcal{T} \text{ uniquely produces } S_N^3\}$[2].

In the figures in this paper, we use big squares to represent tiles placed in the $z = 0$ plane and small squares to represent tiles placed in the $z = 1$ plane. A glue between a $z = 0$ tile and $z = 1$ tile is denoted as a small black disk. Glues between $z = 0$ tiles are denoted as thick lines. Glues between $z = 1$ tiles are denoted as thin lines.

3   Optimal Encoding at Temperature 1

A key problem in algorithmic self-assembly is that of *providing input to a tile assembly system* (e.g., the size of a square, the input to a Turing machine, etc.). In real-world laboratory implementations, as well as theoretical constructions, input to a tile system is typically provided via a (possibly large) collection of "hard-coded" seed tile types that uniquely assemble into a convenient "seed structure," such as a line of tiles that encodes some input value. Unfortunately, in practice, it is more expensive to manufacture different types of tiles than it is to create copies of each tile type. Thus, it is critical to be able to provide input to a tile system using the smallest possible number of hard-coded seed tile types.

---

[2] One subtle difference between our 3D definition of K and the original 2D definition of the tile complexity of an $N \times N$ square, given by Rothemund and Winfree in [7], is that they assume a fully-connected final structure, whereas we do not.

Consider the problem of constructing a tile set that uniquely self-assembles from a single seed tile into a "seed row" that encodes an n-bit binary string, say x. The most straightforward way to do this is to construct a set of n unique tile types that deterministically assemble into a line of tiles of length n, where each tile in the line represents a different bit of x. This simple construction encodes one bit of x per tile, whence its tile complexity is $O(n)$. Note that, in this example, each tile type is an element of a set of size n, yet each tile type encodes only 1 bit of information, instead of the optimal $O(\log n)$ bits. Is there a more efficient encoding construction?

The optimal encoding constructions of Adleman et al. [1], and Soloveichik and Winfree [8] are more efficient methods of encoding input to a tile set. These constructions are based on the idea that each seed row tile type should encode $k = O(\log n)$ bits – instead of a single bit – of x, which means that $O(n/\log n)$ unique tile types suffice to uniquely self-assemble into a seed row that encodes the bits of x. Unfortunately, now the bits of x are no longer conveniently represented in distinct tiles. Fortunately, if k is chosen carefully, then it is possible to use a tile set of size $O(n/\log n)$ to "extract" the bits of x into a more convenient one-bit-per-tile representation, which can be used to seed a binary counter or a Turing machine simulation.

Up until now, all known optimal encoding constructions (e.g., [1,8]) required cooperative binding (that is, temperature $\tau \geq 2$). In what follows, we propose an optimal encoding construction (based on the construction of Soloveichik and Winfree [8]) that works at temperature $\tau = 1$ and is "just barely" 3D, i.e., tiles are only placed in the $z = 0$ and $z = 1$ planes.

3.1   Overview of the Construction

Let $x = x_{n-1}x_{n-2}...x_1x_0$ be the input string, where $x_i \in \{0,1\}$. Let $m = \lceil n/k \rceil$, where k is the smallest integer satisfying $2^k \geq n/\log n$. We write $x = w_0w_1...w_{m-2}w_{m-1}$, where each w_i is a k-bit block. Note that w_0 is padded to the left with leading 0's, if necessary. In the figures in this section, a green tile represents a starting point for some portion of an assembly sequence, and a red tile represents an ending point.

We extract each of the m k-bit blocks within a roughly rectangular region of space of width $O(k)$ and height $O(m)$. We refer to this region of space as a "block extraction region" (or simply "extraction region"). For each $0 \leq i < m$, we extract block w_i in extraction region i. Each extraction region, other than the first and last ones, assembles via a series of gadgets (small groups of tiles that carry out a specific task).

We encode the k bits of a k-bit block as a series of geometric bumps along a path of tiles that makes up the top border of an extraction region. A bump in the $z = 0$ plane represents the bit 0 and a bump in the $z = 1$ plane represents the bit 1. The end result of our construction is an assembly in which each bit of x is encoded in its own bit-bump (see Fig. 10 for an example).

We extract the k-bit blocks in order, starting with the first block w_0, which represents the most significant bits of x. Normally, to carry out this sort of activity at temperature 1 (i.e., to enforce the ordering of tile placements), one has to encode the order of placement directly into the glues of the tiles. However, for our construction, this would essentially mean encoding the number of the block that is being extracted into the glues of the tiles that fill in its extraction region. Unfortunately, doing so, at least in the most straightforward way, results in an increase in tile complexity from the optimal $O(n/\log n)$ to $\Omega(n^2/\log^2 n)$.

Therefore, in our construction, we encode the number of the block that is being extracted as a geometric pattern along a vertical path of tiles that runs along the right side of each extraction region. We call this special geometric pattern the "block number." Then we use a special gadget called the "block-number gadget" to search for this pattern.

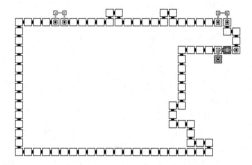

Fig. 1. The perimeter of the first extraction region is hard-coded to self-assemble like this. In this example, the four bumps along the top (from left to right) represent the bits 1, 0, 0 and 1, respectively. The green tile (bottom tile in the penultimate column) is the single seed tile for our entire optimal encoding construction (Color figure online).

Within an extraction region, the block number determines which block gets extracted next. Basically, the path along which the block number is encoded blocks the placement of $m - 1$ special tiles, each of which tries to initiate the extraction of a particular k-bit block. We call these special tiles "extraction tiles." Since the first extraction region is hard-coded (see below), the first block does not have an extraction tile associated with it. Within any given extraction region, exactly one extraction tile will not be blocked. The one extraction tile that is not blocked by the block number gadget will initiate the extraction of the k bits of the block to which it corresponds.

In our construction, we hard-code the assembly of the first and last extraction regions. What this means is that, in each of these extraction regions, a single-tile-wide path assembles the perimeter and then we use $O(1)$ filler tiles to fill in the interior. For this step, it is crucial to first assemble the perimeter of the extraction region and then use the $O(1)$ filler tiles to tile the interior. Note that, if one were to uniquely tile every location in the first (or last) extraction region, then the tile complexity of the construction would be $\Omega(mk)$, which is not optimal. Tiling the perimeter of either the first or last extraction region can

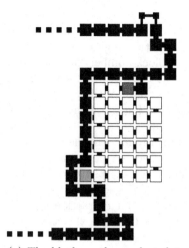

(a) The block-number gadget determines the next block to extract by "searching" for the position of the block number (i.e., the position of the notch in which the red tile is ultimately placed).

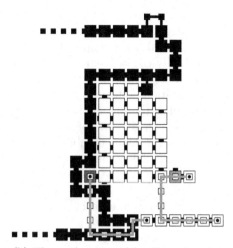

(b) The path initiated by the extraction tile for w_1 "jumps" over the block-number gadget and grows a hook to block a subsequent gadget.

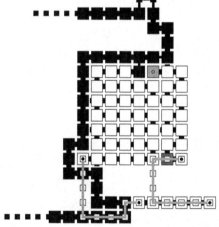

(c) The path initiated by the extraction tile for w_1 continues growing upward and eventually finds the top of the block-number gadget. The upward growth of this path is blocked by a portion of the previous extraction region.

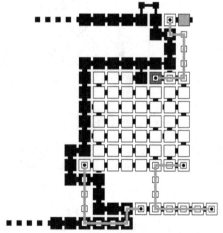

(d) Once at the top of the block-number gadget, the path initiated by the extraction tile for w_1 "jumps" over a portion of the previous extraction region and starts extracting the bits of w_1 along the top of the second extraction region.

Fig. 2. This sequence of figures shows how the position of the block number is found. The black tiles correspond to tiles of the previous extraction region (Color figure online).

be done with $O(m + k)$ unique tile types (see Fig. 1 for the example of the first extraction region).

All extraction regions other than the first and last ones are constructed using a general set of gadgets. In the second extraction region, which is the first generally-constructed extraction region, the block-number gadget determines that w_1 is the next block to be extracted by "searching" for the block number position. When the block number is found, a path of tiles, initiated by the extraction tile for w_1, is allowed to assemble (see Fig. 2 for an example of this process). In general, for extraction region i, for all $1 \leq i < m - 1$, the path along which the block number is encoded geometrically hinders the placement of all extraction tiles that correspond to blocks $w_1, ..., w_{i-1}, w_{i+1}, ..., w_{m-2}$.

Each extraction tile initiates the extraction of the k-bit block to which it corresponds (see Fig. 3). We use a set of "bit-extraction" gadgets to extract a k-bit block into a one-bit-per-bump representation (the bit extraction gadgets are collectively referred to as the "extraction gadget"). Our bit extraction gadgets are basically 3D, temperature 1 versions of the "extract bit" tile types in Fig. 5.7a appears in [8].

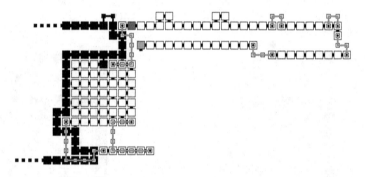

Fig. 3. The bits of the current block are represented as bumps along the top of the extraction region that is currently being assembled.

After a block, say w_i, for $i > 0$, is extracted, the block number is geometrically "incremented", i.e., its position is translated up by a small constant amount (notice the position of the white "hook" at the bottom of Fig. 3). We do this in two phases. First, the current position of the block number is found and then it is incremented and translated. Figure 4 shows how the current position of the block number is detected using a zig-zag path of tiles. Figure 5 shows how the current position of the block number is geometrically incremented.

After the block number has been updated, a series of gadgets geometrically propagate the position of the block number to the right through the remainder of extraction region i so that it is advertised to extraction region $i + 1$. This is shown in Figs. 6 and 7. Technically, we geometrically propagate the block number position through the rest of the extraction region using a series of gadgets.

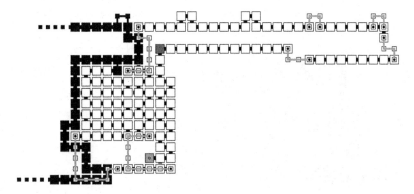

Fig. 4. A path of tiles searches for the block number, represented by a notch in a previous portion of the assembly. The red tile "knows" that it found the position of the block number because it was allowed to be placed (Color figure online).

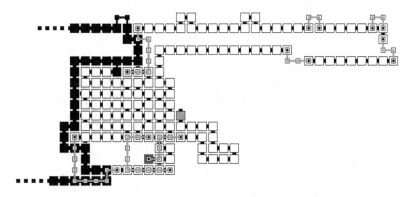

Fig. 5. The block number is geometrically incremented. The green tile "jumps" over the previous gadget that found the position of the block number and grows a hook of tiles to represent the updated block number. Notice that the new hook of tiles is two tiles higher than the previous hook (shown in black), which corresponds to the two rows of tiles that each block takes up in the block-number gadget (Color figure online).

Logically, however, we do this in two phases, which are iterated: "up" propagation and "down" propagation.

The "up" propagation phase grows from the position of the block number up to (and is blocked by) a previous portion of the assembly. This is shown in Fig. 6. The "down" propagation phase grows from the top of the previous (up) propagation phase back down to the position of the block number. The upward growth of each up propagation phase is blocked in the $z = 0$ plane but not in the $z = 1$ plane. However, this is switched for the last up propagation phase. In other words, the last up propagation phase may continue its upward growth, which signals the end of the extraction region, but its $z = 1$ growth is blocked.

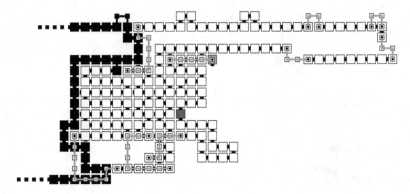

Fig. 6. A series of gadgets geometrically propagate the position of the block number through the rest of the extraction region. This figure shows two of the gadgets. The first one assembles upward until it is blocked by a previous portion of the assembly. The second one assembles horizontally and to the right as it jumps over the top row of the previous gadget.

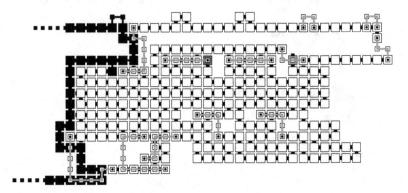

Fig. 7. The position of the block number is propagated through the rest of the extraction region.

In Fig. 8, the last up propagation phase is allowed to continue its upward growth in the $z = 0$ plane.

The last up propagation phase initiates the assembly of a special gadget that fills in the bottom row of the current extraction region before the next extraction region begins. The reason we do this is to ensure that, when the entire extraction process is done (i.e., when all n bits have been extracted into a one-bit-per-bump representation), the bottom row of the assembly is completely filled in. Figure 9 shows an example of how this gadget tiles the remaining perimeter of an extraction region. Note that the tile complexity of this gadget is the size of the perimeter of an extraction region, i.e., $O(m)$.

The final extraction region, like the initial extraction region, is hard-coded to assemble its perimeter via a single-tile-wide path. The tiles that comprise

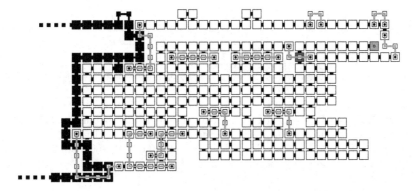

Fig. 8. The last up propagation phase detects when it has reached the end of the extraction region and initiates a perimeter gadget (see Fig. 9) that will fill in the bottom row of the current extraction region before the next extraction region begins.

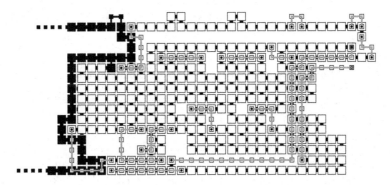

Fig. 9. The bottom row of the extraction region is tiled by a special gadget with $O(m)$ tile complexity. After the bottom row of the extraction region is tiled, the next extraction region is initiated. Notice that the red tile in this figure belongs to the same row of tiles as the red tile in Fig. 1 but the position of the block number has moved up, which means the extraction tile for the next block (in this case, w_2) will be allowed to assemble and all other extraction tiles will be blocked (Color figure online).

the final extraction region "know" to stop the extraction process and possibly initiate the growth of some other logical component of a larger assembly, e.g., a binary counter or a Turing machine simulation in which the extracted bits of x, along the top of each of the m extraction regions, are used as input.

The end result of our optimal encoding construction is a roughly rectangular assembly of tiles with height $O(m)$ and width $O(n)$, where each bit of x is encoded as a bump (either in the $z = 0$ or $z = 1$ plane) along the top of the rectangle, with four "spacer" tiles to the left and right of each bit-bump. Figure 10 shows the result of our optimal encoding construction with four extraction regions.

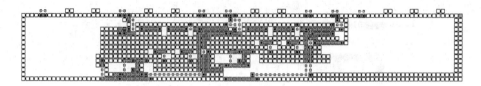

Fig. 10. This is an example of our optimal encoding construction using $n = 16$ and $k = 4$. Note that this does not correspond to an actual instance of our optimal encoding construction because if $n = 16$, then the smallest value of k satisfying $2^k \geq n/\log n$ is $k = 2$. The bit string encoded along the top is 1001001100111000. All of the empty spaces in the $z = 0$ plane are filled in with $O(1)$ filler tiles.

The tile complexity of our construction is $O(n/\log n)$, which is optimal for all algorithmically random values of n.

4   Optimal Self-assembly of Squares at Temperature 1 in 3D

We use our 3D temperature 1 optimal encoding construction to prove the following theorem.

Theorem 1. $K^1_{3DSA}(N) = O\left(\frac{\log N}{\log \log N}\right)$.

Proof. Our proof is constructive. Figure 11a shows how we build an $N \times N$ square using two counters C1 and C2 and two filler regions F1 and F2. Counter C1 is a zig-zag counter whose construction is depicted in Fig. 13. Counter C2 is identical to C1 after a 90-degree clockwise rotation. Each counter is seeded with a value produced by an optimal encoding region (OER for short). The full construction for F1 is depicted in Fig. 12. F2 is a smaller, mirror-image of F1 with minor modifications to properly connect all of the pieces of the square. Both F1 and F2 are essentially squares, except for two hooks needed to stop the horizontal and vertical growths of each filler region, namely, one eight-tile hook encroaching on and another one-tile hook protruding from each filler region (see Fig. 12). These hooks require simple modifications of the OER regions (see Fig. 11b) that are all located in the hard-coded (i.e., first and last) block extracting regions of OER1 and OER2. Note that F1 is also missing a two-tile wide rectangle region on its left that is used up by the vertical connector that initiates the assembly of OER2 immediately after the assembly of C1 terminates. Figure 11c shows the assembly sequence for the whole square, while Fig. 11b zooms in on the region of the square where OER1, F1, OER2 and F2 all interact.

First, we compute the tile complexity of our construction as the sum of the tile complexities of all of the components that make up the $N \times N$ square. Let $n = \lceil \log N \rceil$. If k denotes the smallest integer satisfying $2^k \geq n/\log n$ and m is defined as $\lceil n/k \rceil$, then the tile complexity of each OER is $O(n/\log n)$.

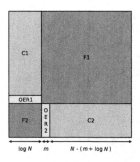

(a) Overall square construction.

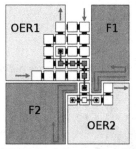

(b) Detail of the region of the square where OER1, F1, OER2 and F2 meet.

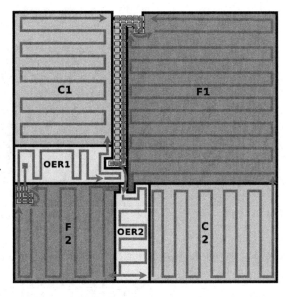

(c) The assembly sequence for the whole square is shown with red arrows starting from the seed tile (the red square located in OER1). The central region in this sub-figure is shown in more detail in Figure 11b to the left.

Fig. 11. Construction of an $N \times N$ square, where m is $O\left(\frac{\log N}{\log \log N}\right)$. The counters C1 and C2 (in medium gray) are identical up to rotation. So are their seed rows, each of which is the output of an optimal extraction region (OER1 and OER2, respectively, in light gray). F1 and F2 (in dark gray) are filler regions (Color figure online).

Furthermore, the tile complexity of each binary counter is $O(1)$. Finally, the tile complexity of each filler region is $O(1)$, since each colored gadget in Fig. 12 has tile complexity $O(1)$. Therefore, the tile complexity of our square construction is dominated by that of the OERs and is therefore $O(\frac{n}{\log n}) = O\left(\frac{\log N}{\log \log N}\right)$.

Second, we need to prove that our tile system is directed and does produce an $N \times N$ square. The assembly sequence depicted in Fig. 11c demonstrates that our tile system uniquely produces a square. To make sure that this square has width N, we need to pick the initial value i of the counters and adjust the size of the filler regions as follows. The width of OER1, C1 and F2 in our construction, and thus also the height of OER2, C2 and F2 is $6n + 4$. The height of OER1, and thus also the width of OER2, is $2m + 7$ (see, for example, Fig. 10). Therefore, the height of C1 (and thus also the width of C2) must be equal to $N - (2m + 7 + 6n + 4) = N - 2\left\lceil \frac{\lceil \log N \rceil}{k} \right\rceil - 6\lceil \log N \rceil - 11$. Let us denote this value by $h(N)$. Our construction of the counters gives us two knobs to control

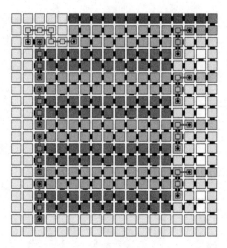

Fig. 12. Detailed construction for the F1 filler region in Fig. 11 (Color figure online)

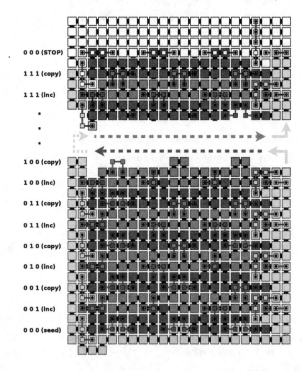

Fig. 13. Detailed construction for the C1 counter in Fig. 11. The gray tiles at the bottom of the counter are part of the optimal extraction region that produces the seed value. The assembly of the counter starts at the orange glue in the bottom-right corner of the figure (Color figure online).

the height of any n-bit counter: the initial value i of the counter and the number r of rooftop rows (the white tiles in Fig. 13), where $r \in \{1, 2, 3, 4\}$. Since each value from i to the final value of the counter $2^n - 1$ (inclusive) takes up four rows of tiles, we must have $\lfloor \frac{h(N)}{4} \rfloor = 2^n - i$ and $r = 1 + h(N) \bmod 4$. Therefore, for both C1 and C2, the initial value of the counter is $2^n - \lfloor \frac{h(N)}{4} \rfloor$. Finally, the correct height and width of F1 are obtained by setting the two knobs described in Fig. 12 to $1 + (2m + 7 + h(N) - 1) \bmod 4$ and $4 - (2m + 7 + h(N) - 2) \bmod 2$, respectively. Similarly, the correct height and width of F2 are obtained by setting the second knob to $4 - (6n + 4) \bmod 2$ and the first knob to $1 + (6n + 4 - 1) \bmod 4$.

The gray tiles in this figure do not belong to F1. They are all added to the $N \times N$ square assembly before F1 starts assembling and they determine the height and width of F1, both of which are adjustable in the following way:

- The height of F1 is always a multiple of four (i.e., the total height of each pink plus red gadget), plus the number of purple rows at the top, which can be hard-coded to any value in $\{1, 3, 4\}$, together with a corresponding increase in the height of the top-left (gray) hook.
- The width of F1 is always a multiple of two (i.e., the width of each pink gadget) plus the width of each orange gadget, which is either three (by deleting the column occupied by the white tiles) or four (as shown).

Therefore, this construction gives us two knobs, namely the number of purple rows and the width of the orange gadget, to assemble filler regions of any height and width, respectively.

5   Conclusion

In this paper, we developed a 3D temperature 1 optimal encoding construction, based on the 2D temperature 2 optimal encoding construction of Soloveichik and Winfree [8]. We then used our construction to answer an open question of Cook, Fu and Schweller [2], namely, we proved that $K^1_{3DSA}(N) = O\left(\frac{\log N}{\log \log N}\right)$, which is the optimal tile complexity for all algorithmically random values of N.

References

1. Adleman, L., Cheng, Q., Goel, A., Huang, M.D.: Running time and program size for self-assembled squares. In: STOC, Huang, pp. 740–748 (2001)
2. Cook, M., Fu, Y., Schweller, R.: Temperature 1 self-assembly: deterministic assembly in 3D and probabilistic assembly in 2D. In: SODA 2011: Proceedings of the 22nd Annual ACM-SIAM Symposium on Discrete Algorithms. SIAM (2011)
3. Doty, D., Patitz, M.J., Summers, S.M.: Limitations of self-assembly at temperature 1. Theor. Comput. Sci. **412**, 145–158 (2011)
4. Li, M., Vitányi, P.: An Introduction to Kolmogorov Complexity and its Applications, 3rd edn. Springer, New York (2008)

5. Manuch, J., Stacho, L., Stoll, C.: Two lower bounds for self-assemblies at temperature 1. J. Comput. Biol. **17**(6), 841–852 (2010)
6. Presburger, Mojżesz: Über die vollständigkeit eines gewissen systems der arithmetik ganzer zahlen, welchem die addition als einzige operation hervortritt. In: Compte-Rendus du Premier Congrès des Mathématiciens des Pays Slaves, Warsaw, pp. 92–101 (1930)
7. Rothemund, P. W., Winfree, E.: The program-size complexity of self-assembled squares (extended abstract). In: STOC 2000: Proceedings of the Thirty-second Annual ACM Symposium on Theory of Computing, pp. 459–468 (2000)
8. Soloveichik, D., Winfree, E.: Complexity of self-assembled shapes. SIAM J. Comput. **36**(6), 1544–1569 (2007)
9. Wang, H.: Proving theorems by pattern recognition - II. Bell Syst. Tech. J. XL **40**(1), 1–41 (1961)
10. Winfree, E.: Algorithmic self-assembly of DNA. Ph.D. thesis, California Institute of Technology, June 1998

Flipping Tiles: Concentration Independent Coin Flips in Tile Self-Assembly

Cameron T. Chalk$^{(\boxtimes)}$, Bin Fu, Alejandro Huerta, Mario A. Maldonado, Eric Martinez, Robert T. Schweller, and Tim Wylie

University of Texas Rio Grande Valley, Edinburg, TX 78539-2999, USA
{cameron.chalk01,bin.fu,alejandro.huerta02,eric.m.martinez02,
robert.schweller,timothy.wylie}@utrgv.edu

Abstract. In this paper we introduce the *robust coin flip* problem in which one must design an abstract tile assembly system (aTAM system) whose terminal assemblies can be partitioned such that the final assembly lies within either partition with exactly probability 1/2, regardless of what relative concentration assignment is given to the tile types of the system. We show that robust coin flipping is possible within the aTAM, and that such systems can guarantee a worst case $\mathcal{O}(1)$ space usage. As an application, we then combine our coin-flip system with the result of Chandran, Gopalkrishnan, and Reif [3] to show that for any positive integer n, there exists a $\mathcal{O}(\log n)$ tile system that assembles a constant-width linear assembly of expected length n that works for all concentration assignments. We accompany our primary construction with variants that show trade-offs in space complexity, initial seed size, temperature, tile complexity, bias, and extensibility, and also prove some negative results. Further, we consider the harder scenario in which tile concentrations change arbitrarily at each assembly step and show that while this is not solvable in the aTAM, this version of the problem can be solved by more exotic tile assembly models from the literature.

1 Introduction

Self-assembly is the process by which local interactivity among unorganized, autonomous units results in their amalgamation into compounds. One of the premiere models for studying the theoretical possibilities of self-assembly is the *abstract tile assembly model* (aTAM) [22] in which system monomers are 4-sided Wang tiles that attach to a growing seed assembly whenever matching glues present a sufficient bonding strength. The motivation for studying the aTAM stems from the feasibility of a nanoscale DNA implementation [12], along with the universal computational power of the model [19], which permits many features including *algorithmic* self-assembly of general shapes [20], and more [8,17].

C.T. Chalk, A. Huerta, M.A. Maldonado, E. Martinez and R.T. Schweller—Research supported in part by National Science Foundation Grant CCF-1117672.

B. Fu—Research supported by National Science Foundation Early Career Award 0845376.

© Springer International Publishing Switzerland 2015
A. Phillips and P. Yin (Eds.): DNA 2015, LNCS 9211, pp. 87–103, 2015.
DOI: 10.1007/978-3-319-21999-8_6

A promising new direction in self-assembly is the consideration of *randomized* self-assembly systems. In randomized self-assembly (a.k.a. nondeterministic self-assembly), assembly growth is dictated by nondeterministic, competing assembly paths yielding a probability distribution on a set of final, terminal assemblies. By careful design of tile-sets and the relative concentration distributions of these tiles, a number of new functionalities and efficiencies have been achieved that are provably impossible without this non-determinism. For example, by precisely setting the concentration values of a generic set of tile species, arbitrarily complex strings of bits can be *programmed* into the system to achieve a specific shape with high probability [9,15]. Alternately, if the concentration of the system is assumed to be fixed at a uniform distribution, randomization still provides for efficient expected growth of linear assemblies [3] and low-error computation at temperature-1 [6]. Even in the case where concentrations are unknown, randomized self-assembly can build certain classes of shapes without error in a provably more efficient manner than without randomization [2].

Motivated by the power of randomized self-assembly, along with the potential for even greater future impact, we focus on the development of the most fundamental randomization primitive: the *robust* generation of a uniform random bit. In particular, we introduce the problem of self-assembling a uniformly random bit within $\mathcal{O}(1)$ space that is guaranteed to work for all possible concentration distributions. We define a tile system to be a *coin flip* system, with respect to some tile concentration distribution, if the terminal assemblies of the system can be partitioned such that each partition has exactly probability 1/2 of assembling. We say a system is a *robust coin flip* system if such a partition exists that guarantees 1/2 probability for all possible tile concentration distributions. By designing systems that flip a fair coin for all possible (adversarially chosen) concentration distributions, we achieve an intrinsically fair coin-flipping system that is robust to the experimental realities of imprecise quantity measurements. Such intrinsically fair systems may further allow for increased scalability of randomized self-assembly systems in scenarios where exact concentrations of species are either unknown or intractable to predict at successive assembly stages.

Our results. Our primary result is an aTAM construction that constitutes a robust fair coin flip system which completes in a guaranteed $\mathcal{O}(1)$ space. We apply our robust coin-flip construction to the result of Chandran, Gopalkrishnan, and Reif [3] to show that for any positive integer n, there exists a $\mathcal{O}(\log n)$ tile system that assembles a constant width-4 linear assembly of expected length n that works for all concentration assignments. We accompany this result with a proof that such concentration independent assembly of width-1 assemblies is not possible with fewer than n tile types. We further accompany our main coin-flip construction with variant constructions that provide trade-offs among standard aTAM metrics such as space, tile complexity, and temperature, as well as new metrics such as coin bias, and the *extensibility* of the system, which is the maximum number of distinct locations a single assembly of the system can add a tile. We show that 1-extensible systems, while computationally universal, cannot robustly coin-flip in bounded space without incurring a bias, but can robustly

Table 1. τ represents the temperature of the system, $|\sigma|$ represents the number of tiles in the seed assembly, and k-ext denotes the extensibility of the system. p represents the largest disparity in relative tile concentration between any pair of tile types in the system for a given concentration distribution.

Robust Coin Flip in the aTAM					
Space	Bias	τ	$|\sigma|$	k-ext	Theorem
$\mathcal{O}(1)$	-	1	7	2	1
$\mathcal{O}(1)$	-	2	1	2	2
unbounded	-	2	1	1	4
c	$<p^{(c/2)+1}$	2	1	1	5

Unstable Concentrations Robust Coin Flip					
Model	Space	Bias	τ	$|\sigma|$	Theorem
neg-aTAM	$\mathcal{O}(1)$	-	1	2	9
neg-hTAM	$\mathcal{O}(1)$	-	1	1	9
polyTAM	$\mathcal{O}(1)$	-	2	3	9
GTAM	$\mathcal{O}(1)$	-	1	2	9

coin-flip in bounded expected space. We also consider the more extreme model in which concentrations may change adversarially at each assembly step. We show that the aTAM cannot robustly coin flip in bounded space within this model, but a number of more exotic extensions of the aTAM from the literature are able to robustly coin flip in $\mathcal{O}(1)$ space. We summarize our results in Table 1. The problem of self-assembling random bits has been considered before [11], but their technique, and in fact almost all randomized techniques to date, do not work when arbitrary concentrations are considered.

2   Definitions and Model

2.1   Tiles, Assemblies, and Tile Systems

Consider some alphabet of glue types Π. A tile is a unit square with 4 edges each assigned some glue type from Π. Further, each glue type $g \in \Pi$ has some non-negative integer strength $str(g)$. Each tile may be assigned a finite length string label, e.g., "black","white", "0", or "1". Further, for simplicity, we assume each tile center is located at a pixel $p = (p1,p2) \in \mathbb{Z}^2$. For a given tile t, we denote the tile center of t as its position. As notation, we denote the set of all tiles that constitute all translations of the tiles in a set T as the set T^*. An assembly is a set of tiles each assigned unique coordinates in $\mathbb{Z}^2$. For a given assembly α, define the bond graph G_α to be the weighted graph in which each element of α is a vertex, and each edge weight between tiles is $str(g)$ if the tiles share an overlapping glue g, and 0 otherwise. An assembly α is said to be τ-stable for a positive integer τ if the bond graph G_α has min-cut at least τ, and τ-unstable otherwise. A tile system is an ordered triple $\Gamma = (T, \sigma, \tau)$ where T is a set of tiles called the tile set (we refer to elements of T as tile types), σ is an assembly called the seed and τ is a positive integer called the temperature. When considering a tile a that is some translation of an element of a tile set T, we will use the term tile type of a to reference the element of T that a is a translation of. Assembly proceeds by growing from assembly σ by any sequence of single tile attachments from T so long as each tile attachment connects with strength at least τ. Formally, we define what can be built in this fashion as the set of producible assemblies:

Definition 1 (Producibility). *For a given tile system $\Gamma = (T, \sigma, \tau)$, the set of **producible assemblies** for system Γ, PROD$_\Gamma$, is defined recursively:*

- *(Base) $\sigma \in$ PROD$_\Gamma$*
- *(Recursion) For any $A \in$ PROD$_\Gamma$ and $b \in T^*$ such that $C = A \cup \{b\}$ is τ-stable, then $C \in$ PROD$_\Gamma$.*

As additional notation, we say $A \to_1^\Gamma B$ if A may grow into B through a single tile attachment, and we say $A \to^\Gamma B$ if A can grow into B through 0 or more tile attachments. An **assembly sequence** for a tile system Γ is a sequence (finite or infinite) $\overrightarrow{\alpha} = \langle \alpha_1, \alpha_2, \dots \rangle$ in which $\alpha_1 = \sigma$, each α_{i+1} is a single-tile extension of α_i, and each α_i is τ-stable. The **frontier** of an assembly α, written as $F(\alpha, \Gamma)$, is a partial function that maps an assembly α and a tile system Γ to a set of tiles $\{t \in T^* | \alpha \cup \{t\} \in$ PROD$_\Gamma \wedge t \notin \alpha\}$. We further define TERM$_\Gamma$ to be the subset of PROD$_\Gamma$ consisting only of assemblies for which no further tile in T may attach.

Definition 2 (Finiteness and Space). *For a given tile assembly system $\Gamma = (T, \sigma, \tau)$, we say Γ is **finite** iff $\forall \sigma \in$ PROD$_\Gamma, \exists \alpha \in$ TERM$_\Gamma : \sigma \to^\Gamma \alpha$. That is, each producible assembly has a growth path ending in a finite, terminal assembly. If Γ is not finite, we say it is **infinite**. Define the **space of an assembly** α as $|\alpha|$. Let the **space of a tile assembly system** be defined as the $\max\limits_{\alpha \in TERM_\Gamma} |\alpha|$ iff Γ is finite. If Γ is infinite, let **space** remain undefined. Note that a finite system may have infinite/unbounded space.*

Definition 3 (Extensibility). *Consider a tile assembly system $\Gamma = (T, \sigma, \tau)$, and assembly $\alpha \in$ PROD$_\Gamma$. We denote the set of all locations at which a tile may stably attach to α as L_α. More formally, $L_\alpha = \{p_t | t \in F(\alpha, \Gamma)\}$. We say a tile system Γ is k-**extensible** iff $\forall \alpha \in$ PROD$_\Gamma, |L_\alpha| \leq k$. Informally, a tile assembly system is k-extensible iff at any point in the assembly process, the assembly can only grow in at most k locations.*

2.2 Probability in Tile Assembly

We use the definition of probabilistic assembly presented in [1,3,6,9,15]. Let P be a function denoting a **concentration distribution** over a tileset T representing the concentrations of each tile type with the restrictions $\forall t \in T, P(t) > 0$ and $\sum\limits_{t \in T} P(t) = 1$. For a tile t, we sometimes refer to $P(t)$ as the **concentration** of t. Using a concentration distribution, we can consider probabilities for certain events in the system. To study probabilistic assembly, we can consider the assembly process as a Markov chain where each producible assembly is a state and transitions occur with non-zero probability from assembly A to each B whenever $A \to_1^\Gamma B$. For each B that satisfies $A \to_1^\Gamma B$, let $t_{A \to B}$ denote the tile in T whose translation is added to A to get B. The transition probability from A to B is defined to be

$$TRANS(A, B) = \frac{P(t_{A \to B})}{\sum_{\{C | A \to_1^\Gamma C\}} P(t_{A \to C})} \quad (1)$$

The probability that a tile system Γ terminally assembles an assembly A is defined to be the probability that the Markov chain ends in state A. For each $A \in \text{TERM}_\Gamma$, let $\text{PROB}_{\Gamma \to A}^P$ denote the probability that Γ terminally assembles A with respect to concentration distribution P.

Definition 4 (Expected Space). *For a given finite tile system $\Gamma = (T, \sigma, \tau)$, let the expected space of Γ relative to a concentration distribution P be defined as*

$$EXPECTEDSPACE_\Gamma = \sum_{\alpha \in TERM_\Gamma} |\alpha| \cdot PROB_{\Gamma \to \alpha}^P \quad (2)$$

Definition 5 (Coin Flipping). *We consider a finite tile system Γ a **coin flip tile system with bias** b with respect to a concentration distribution P for some $b \in \mathbb{R}$ iff the set of terminal assemblies in $PROD_\Gamma$ is partitionable into two sets X and Y such that* $\left| \sum_{x \in X} PROB_{\Gamma \to x}^P - \sum_{y \in Y} PROB_{\Gamma \to y}^P \right| \leq 2b$. *A **fair coin flip tile system** is a coin flip tile system with bias 0. We consider a finite tile system Γ a **robust coin flip tile system with bias** b iff the set of terminal assemblies in $PROD_\Gamma$ is partitionable into two sets X and Y such that* $\left| \sum_{x \in X} PROB_{\Gamma \to x}^C - \sum_{y \in Y} PROB_{\Gamma \to y}^C \right| \leq 2b$ *for all concentration distributions C. A **robust fair coin flip tile system** is a robust coin flip tile system with bias 0.*

3   Robust Fair Coin Flipping in the aTAM

In this section we show systems capable of robust fair coin flips in the aTAM. Figure 1 shows a simple fair coin flip aTAM system for the uniform concentration distribution. To solve this problem for arbitrary concentration distributions, more involved techniques are required.

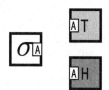

Theorem 1. *There exists a $\mathcal{O}(1)$ space 2-extensible robust fair coin flip tile system $\Gamma = (T, \sigma, 1)$ in the aTAM with $|\sigma| = 7$.*

Fig. 1. A non-robust fair coin flip for the uniform concentration distribution.

Proof. To show this we present a tile system $\Gamma = (T, \sigma, 1)$ in which two terminal states exist and are equiprobable for all concentration distributions P. $|T| = 9$ and σ contains 7 tiles. The system terminates nondeterministically and contains either 2 h tiles and 1 t tile or 2 t tiles and 1 h tile. The system leverages any difference in tile concentrations

Fig. 2. Shown are the σ, h, and t tiles on the left, and the terminal states of the assembly system representing heads and tails. A, B and C glues are strength 1. Non-matching glues have 0 strength.

between h and t by ensuring that placement of a t tile increases the probability of terminating in an assembly containing $2h$ tiles and vice versa. A graphical representation of σ, the h and t tiles, and terminal states of the assembly system is shown in Fig. 2. Without loss of generality, assume the leftmost bottom tile in σ sits at position $(0,0)$. We will refer to each producible assembly sans σ by the labels of the tiles in positions $(1,1), (2,1)$ and $(3,1)$ as such: $\_t, h\_\_, \_ht$ and so forth. We now show that $\text{PROB}^P_{\Gamma \to hht} = \dfrac{1}{2}$ for all concentration distributions P. Let c_h be the concentration of the tile labeled h and c_t be the concentration of the tile labeled t, then

$$
\begin{aligned}
\text{PROB}^P_{\Gamma \to hht} &= TRANS(\sigma, \_t) \cdot TRANS(\_t, \_ht) \cdot TRANS(\_ht, hht) \\
&+ TRANS(\sigma, \_t) \cdot TRANS(\_t, h\_t) \cdot TRANS(h\_t, hht) \\
&+ TRANS(\sigma, h\_\_) \cdot TRANS(h\_\_, h\_t) \cdot TRANS(h\_t, hht) \\
&= \frac{c_t}{c_t + c_h} \cdot \frac{c_h}{c_h + c_h} \cdot \frac{c_h}{c_h} + \frac{c_t}{c_t + c_h} \cdot \frac{c_h}{c_h + c_h} \cdot \frac{c_h}{c_t + c_h} \\
&+ \frac{c_h}{c_t + c_h} \cdot \frac{c_t}{c_t + c_t} \cdot \frac{c_h}{c_t + c_h} \\
&= \frac{c_t{}^2 + 2c_t c_h + c_h{}^2}{2c_t{}^2 + 4c_t c_h + 2c_h{}^2} = \frac{1}{2}.
\end{aligned}
$$

$\square$

3.1 Extension to a Single-Seed

A common constraint in the aTAM is that σ contains only one tile. Thus, no seed structure must be formed prior to the self-assembly process. The construction shown in Fig. 3 addresses this constraint and works in a similar fashion as the construction in Theorem 1. Note that this system requires $\tau = 2$.

Theorem 2. *There exists a $\mathcal{O}(1)$ space 2-extensible robust fair coin flip tile system $\Gamma = (T, \sigma, 2)$ in the aTAM with $|\sigma| = 1$.*

Proof. Our tile set is shown in Fig. 3. Without loss of generality, assume σ sits at position $(0,0)$. Until the tile labeled S (see Fig. 3) is placed, the assembly process is deterministic. Upon attachment of S, cooperative binding locations

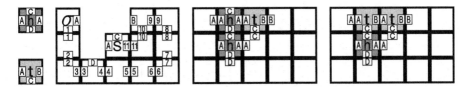

Fig. 3. T is shown. Our seed, labeled σ, begins a deterministic attachment process ending with the placement of the tile labeled S. Glues labeled $\{1, 2, 3, \ldots, 11\}$ are of strength 2. Glues labeled $\{A, B, C, D\}$ are of strength 1, ensuring that the nondeterministic attachments of tiles h and t do not begin until the cooperative binding locations are opened by placement of the tile labeled S. The nondeterministic sequence of attachments following the placement of S is similar to that of Theorem 1.

allow the attachment of tiles h and t nondeterministically. We denote the assemblies following the placement of S similarly to the proof of Theorem 1. We refer to assemblies containing tile S by the labels of tiles in positions $(1, -1), (1, 0)$ and $(2, 0)$ as $\_t, \_h, \_ht$ and so forth. Reflecting the analysis shown in Theorem 1, we have $\text{PROB}^P_{\Gamma \to hht} = .5$ for all concentration distributions P, which implies $\text{PROB}^P_{\Gamma \to htt} = .5$ as there are two terminal assemblies $\square$.

3.2   1-Extensible Coin Flipping

The previous sections showcase 2-extensible solutions to the robust fair coin flip problem. A natural question follows: is there a 1-extensible solution? Theorem 3 shows that there is no $\mathcal{O}(1)$ space solution in the aTAM. Using algorithms based on John von Neumann's randomness extractor [21] we can achieve an unbounded space robust fair coin flip system (Theorem 4) as well as a $\mathcal{O}(1)$ space construction which incurs a small bias (Theorem 5).

Theorem 3. *There does not exist a $\mathcal{O}(1)$ space 1-extensible robust fair coin flip tile system in the aTAM.*

Proof. We prove this by contradiction. Assume that there exists a $\mathcal{O}(1)$ space 1-extensible robust fair coin flip aTAM tile system $\Gamma = (T, \sigma, \tau)$. We now specify a concentration distribution for m tiles in T that contradicts this claim. Assume that Γ generates assemblies of size at most h. Consider a series of phases $p_1, \ldots, p_n$ such that p_{i+1} is derived from p_i by the attachment of the tile in the frontier of p_i with the largest concentration. Select a parameter $t = 10mn^3$, and let $c_1 = 1$ and $c_{i+1} = tc_i$ for $i = 1, \ldots, m - 1$. Let the concentration for each $t_i \in T$ be $\frac{c_i}{c_1 + c_2 + \cdots + c_m}$.

For each assembly p_i, let $q_{i_1}, \ldots, q_{i_u}$ be the set of tile types in the frontier of p_i listed in increasing order by their concentrations. Let c_{i_u} denote the concentration of tile type q_{i_u}. With probability $\frac{c_{i_u}}{c_{i_1} + \cdots + c_{i_u}}$, tile type q_{i_u} is attached. We have

$$\frac{c_{i_u}}{c_{i_1} + \cdots + c_{i_u}} \geq \frac{1}{\frac{(u-1)c_{i_{u-1}}}{c_{i_u}} + 1} \tag{3}$$

$$\geq \frac{1}{\frac{(u-1)}{t} + 1} \geq \frac{1}{\frac{m}{t} + 1} \tag{4}$$

$$\geq \frac{1}{\frac{1}{10n^3} + 1}. \tag{5}$$

Therefore, with probability at least

$$\left(\frac{1}{\frac{1}{10n^3} + 1}\right)^n \geq \left(\frac{1}{\frac{1}{10n^3} + 1}\right)^{10n^3 \cdot \frac{1}{10n^2}} \tag{6}$$

$$\geq \left(\frac{1}{e}\right)^{\frac{1}{10n^2}} > 0.6 \tag{7}$$

we follow the sequence $p_1, \ldots, p_n$ to generate an assembly. This is a contradiction. Note that we use the facts that $(1 + \frac{1}{x})^x$ is an increasing function for all real $x > 1$, and $\lim_{x \to +\infty}(1 + \frac{1}{x})^x = e \approx 2.17828$. $\qquad\square$

In response to Theorem 3, we give a 1-extensible aTAM system capable of robust fair coin flips in unbounded space in Theorem 4. In 1951, John von Neumann gave a simple method for extracting a fair coin from a biased one [21]. We show two algorithms based on the Von Neumann extractor. Algorithm 1 uses an unbounded number of *rounds* to extract a fair coin flip. We use Algorithm 1 to show that a *fair coin flip extractor* can be implemented in the aTAM to achieve an unbounded space, 1-extensible, robust coin flip tile system. We extend this method in Algorithm 2 to create a *bounded fair coin flip extractor* by adding a parameter k which controls the maximum number of rounds allowed. This is a *bounded coin flip extractor* that is implemented in the aTAM and achieves $\mathcal{O}(1)$ space, is 1-extensible, and is a robust coin flip tile system with bounded bias.

Algorithm 1. Unbounded

```
1: procedure UNBOUNDEDFCFE(h, t)
2:     coin = {heads, tails}
3:     pdist = {h, t}
4:     repeat
5:         flip_1 ← flip(coin, pdist)
6:         flip_2 ← flip(coin, pdist)
7:     until flip_1 ≠ flip_2
8:     return flip_2
9: end procedure
```

Algorithm 2. Bounded

```
1: procedure BOUNDEDFCFE(h, t, k)
2:     coin = {heads, tails}
3:     pdist = {h, t}
4:     round ← 1
5:     while round ≤ k do
6:         flip_1 ← flip(coin, pdist)
7:         flip_2 ← flip(coin, pdist)
8:         if flip_1 ≠ flip_2 then
9:             return flip_2
10:        end if
11:        round ← round + 1
12:    end while
13:    return flip(coin, h, t)
14: end procedure
```

(a) Tile set that makes two nondeterministic *flips* corresponding to the two calls to the *flip* function in 1.

(b) Tile set that checks the result of the two flips and possibly starts another round if a fair bit has not been achieved. A *HEADS* or *TAILS* tile is placed if a fair bit has been achieved.

Fig. 4. The tile labeled S is the seed of the tile assembly system and the temperature is 2. The strength of the glues are as follows: str(0)=1, str(1)=1, str(A)=2, str(B)=2, str(C)=1, str(D)=1, str(F)=1, str(G)=2, str(R)=2, and str(R')=2.

We now describe our 1-extensible aTAM tile system that implements Algorithm 1. In Algorithm 1, a coin is a set of cardinality 2 with possible values *heads* and *tails*. *flip* is a function that selects and returns a *heads* or *tails* value based on the probabilities h and t, where $h, t \in (0,1)$ and $h + t = 1$. In our construction, calls to the *flip* function are carried out by a non-deterministic competition for attachment between a *0* tile and a *1* tile. Aside from calls to the *flip* function, the rest of the algorithm can be implemented by deterministic tile placements. Figure 4 gives the tile set used in the construction. Consider all tiles labeled H as *HEADS* tiles and all tiles labeled T as *TAILS* tiles where their placement implies the returning of heads and tails, respectively. Consider all tiles labeled E as *ERR* tiles. The set of tiles in Fig. 4(a) starts the process and makes two non-deterministic placements of a *1* tile or a *0* tile. The set of tiles in Fig. 4(b) checks the result of the two flips. If the order of the flips, starting from the left, is *10*, it outputs a *HEADS* tile. If the order of the flips is *01*, it outputs a *TAILS* tile. Otherwise, it outputs an *ERR* tile, which starts another loop. Figure 5 shows examples of assemblies that can grow in Round 1 of the algorithm. This construction yields Theorem 4. The full analysis of this construction is omitted in this version due to space.

Theorem 4. *There exists a 1-extensible, robust coin flip tile system in the aTAM. The tile system achieves $\mathcal{O}(1/pq)$ expected space, where p and q denote the relative concentrations of the two tiles with the largest difference in concentration for a given concentration distribution.*

We now extend Algorithm 1 by adding a parameter k, which controls the maximum number of rounds allowed (Algorithm 2). This *bounded fair coin flip extractor* can be implemented in the aTAM to achieve a $\mathcal{O}(1)$ space, 1-extensible, robust coin flip tile system with bounded bias. The bounded k-rounds can be controlled by the implementation of a 1-extensible version the the aTAM counter construction from [5] for a desired base, leading to a tradeoff in bias, space, and tile complexity. We state the primary tradeoff in Theorem 5 between space and bias, and omit the tradeoff in tile complexity in this version, as well as construction details and analysis.

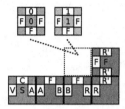

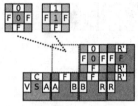

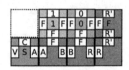

(a) An assembly with two possible choices for the next attachment corresponding to the first flip in the algorithm.

(b) Without loss of generality, this shows possible choices for the second flip of the algorithm after the first has been chosen.

(c) A *0* tile and a *1* tile have been placed for the first and second flip, respectively. From Algorithm 1, this will return a heads.

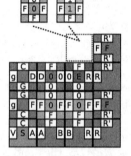

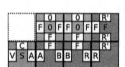

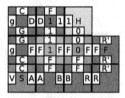

(d) Two *0 tiles* were placed for the first two flips. From Algorithm 1, the system must perform another round.

(e) An assembly where the first round of the algorithm failed to generate a bit and proceeds to start a new round.

(f) An assembly where the first round of the algorithm was a valid flip and it generates a heads.

Fig. 5. A sample of producible assemblies for round 1

Theorem 5. *There exists a c space 1-extensible robust coin flip tile system in the aTAM with bias less than $p^{(c/2)+1}$, where p denotes the larger relative concentration of the pair of tiles with the largest difference in concentration for a given concentration distribution.*

4   Robust Simulation of Randomized Linear Assemblies

As an application of the primitive shown in Theorem 2, we show that a class of randomized linear aTAM tile assembly systems can be simulated in a concentration robust manner with a minor scale factor.

We first briefly describe a scale (m, n)-simulation of a given tile system, based on the block replacement schemes of [4]. Consider an aTAM system $\Gamma = (T, \sigma, \tau)$ and a proposed simulator system $\Gamma' = (T', \sigma', \tau')$. Now consider the mapping from TERM_Γ to $\text{TERM}_{\Gamma'}$ obtained by replacing each tile in an assembly $A \in \text{TERM}_\Gamma$ with a rectangular $m \times n$ block of tiles over U, according to some fixed $m \times n$

block mapping R. If there exists such a mapping M from TERM_Γ to $\text{TERM}_{\Gamma'}$ that is bijective, then we say that Γ' simulates the production of Γ at scale factor (m, n). Further, we say that Γ robustly simulates Γ' for concentration distribution P if for all terminal assemblies $A \in \text{TERM}_\Gamma$, $\text{PROB}_{\Gamma \to A}^P = \text{PROB}_{\Gamma' \to M(A)}^C$ for all concentration distributions C over T', i.e., Γ' produces terminal assemblies with probability independent of concentration assignment, and with exactly the same probability distribution as the concentration dependent system it simulates.

We now define a class of linear assembly systems for which we can construct robust, concentration independent simulations.

Definition 6 (Unidirectional Two-Choice Linear Assembly Systems).
A tile system Γ is a unidirectional two-choice linear assembly system iff:

1. *Γ is 1-extensible,*
2. *$\forall \alpha \in PROD_\Gamma, |F(\alpha, \Gamma)| \leq 2$,*
3. *$\forall \beta \in PROD_\Gamma, \beta$ is a $1 \times n$ line for some $n \in \mathbb{N}$.*

Theorem 6. *For any unidirectional two-choice linear assembly system $\Gamma = (T, \sigma, \tau)$ in the aTAM, there is an aTAM system $\Gamma_s = (T', \sigma', \tau')$ that robustly simulates Γ for the uniform concentration distribution at scale factor 5×4; further, $|T'| = c|T|$ for some constant c.*

Proof. Let $\Gamma = (T, \sigma, \tau)$ be a unidirectional two-choice linear assembly system. Define an *undecided assembly* to be any assembly $\alpha \in PROD_\Gamma$ such that $|F(\alpha, \Gamma)| = 2$. For each undecided assembly, we will construct a gadget utilizing the technique in Theorem 2. We call the two tiles of an undecided assembly's frontier h and t. Consider $\alpha_h = \alpha \cup h$ and $\alpha_t = \alpha \cup t$. We simulate Γ in reference to a uniform concentration distribution, so α transitions to α_h with probability .5 and to α_t with probability .5. Figure 6 shows an example of utilizing a 5×4 gadget in Γ_s to simulate the transition from α to α_h or α_t. By application of Theorem 2, the gadget will grow into one of two possible states with probability .5 for any concentration distribution. By chaining the gadgets together we can robustly simulate the nondeterministic attachments in Γ. Each tile is simulated by a 5×4 block of tiles, therefore $|T'| = c|T|$ for some constant c. □

As a corollary to Theorem 6, we can create a tile system to build an expected length n assembly for all concentration distributions with $\mathcal{O}(\log n)$ tile complexity. First, we will prove that there is no aTAM tile system which generates linear (width-1) assemblies of expected length n for all concentration distributions ([3] showed that this is possible for the uniform concentration distribution).

Theorem 7. *There is no aTAM tile system to generate an assembly of width-1 and expected length n for all concentration distributions with less than n tile complexity.*

Proof. Towards a contradiction, assume a self-assembly system can generate a linear assembly with expected length n and uses at most $k < n$ tiles. There is at least one assembly S that is of length at least n. Let $S = t_1 \cdots t_{i-1} t_i \cdots t_m t_i ...,$

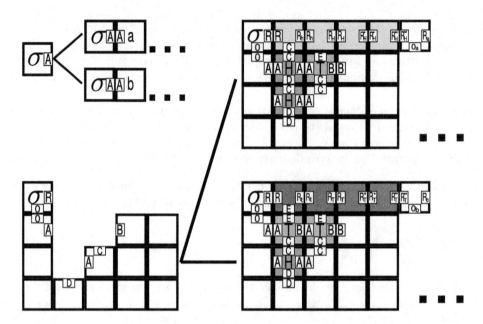

Fig. 6. A simulation of one non-deterministic linear tile attachment. Each non-determinstic attachment will require a 5×4 robust coin flip gadget shown in Fig. 3. The assembly may continue after simulating a non-determinstic attachment by building another 5×4 robust coin flip gadget, building a deterministic 5×4 block, or terminating.

where $t_i \cdots t_m t_i$ is the first cycle that appears in S since there are less than n tiles. We define the concentration of the types of tiles as follows:

Let $c_1 = 1$, $c_j = c_{j-1}/n^{100}$ for $j = 1, ..., k$. The concentration of each type t_i is $\frac{c_i}{c_1+c_2+\cdots+c_k}$. Therefore, with probability at least $(\frac{1}{1+\frac{1}{n^{99}}})^n$, the assembly $t_1 \cdots t_{i-1} t_i \cdots t_m t_i$, or one at least as long, will be generated. With probability at least $(\frac{1}{1+\frac{1}{n^{99}}})^{n^3} > 0.9$, an assembly at least as long as $t_1 \cdots t_{i-1}(t_i \cdots t_m t_i)^{n^2} \ldots$ will be generated, which has length at least n^2. This contradicts the assumption that the expected length is n. □

We now contrast the width-1 impossibility result of Theorem 7 with a result showing that width-4 linear assemblies do allow for efficient growth to expected length n in a concentration independent manner. To achieve this, we apply Theorem 6 to the unidirectional two-choice linear assembly system presented in [3], which yields the following result.

Corollary 1. *There exists an aTAM tile system $\Gamma = (T, \sigma, \tau)$ which terminates in a width-4 expected length n assembly for all concentration distributions. $|T| = \mathcal{O}(\log n)$.*

Proof. Let m be $\lfloor \frac{n}{5} \rfloor$. Consider $\Gamma = (T, \sigma, \tau)$ to be a robust simulation at scale factor 5×4 of a unidirectional two-choice linear assembly system that terminates in an expected length m linear assembly using $O(\log m)$ tile types. Note that such a unidirectional two-choice linear assembly system exists as shown in [3] and can be robustly simulated as shown by Theorem 6. If $5m = n$, then Γ terminates in an expected length n assembly with width-4; otherwise, we add $n \mod 5$ length deterministically. Since our scale factor is constant, $|T| = O(\log n)$. $\qquad \square$

5   Robust Fair Coins with Unstable Concentrations

As an extension to the idea of concentration independent solutions outlined in this paper, we consider an adversarial model wherein the concentration distribution of tiles changes during each stage of the assembly process; in other words, the concentrations are unstable.

Definition 7 (Unstable Concentrations Robust Fair Coin Flip). *Let an unstable concentration distribution P be a function mapping $z \in \mathbb{Z}^+$ to concentration distributions over a tile set T. Let P_i denote $P(i)$. For each B that satisfies $A \rightarrow_1^\Gamma B$, let $t_{A \rightarrow B}$ denote the tile in T whose translation is added to A to get B. The transition probability from A to B is defined to be*

$$TRANS(A, B) = \frac{P_{|A|}(t_{A \rightarrow B})}{\sum_{\{C | A \rightarrow_1^\Gamma C\}} P_{|A|}(t_{A \rightarrow C})} \tag{8}$$

*We consider a finite tile system Γ an **unstable concentrations robust fair coin flip** iff the set of terminal assemblies in $PROD_\Gamma$ is partitionable into two sets X and Y such that $\sum_{x \in X} PROB_{\Gamma \rightarrow x}^C = \sum_{y \in Y} PROB_{\Gamma \rightarrow y}^C$ for all unstable concentration distributions C.*

We now prove that there is no unstable concentration robust fair coin flip system in the aTAM. First, we state and prove a lemma that will be useful in our proof.

Lemma 1. *For any producible assembly $A \in PROD_\Gamma$ and any tile type $t \in T$, there exists another assembly A^* such that for any sequence of assemblies $\langle A_0 = A, A_1, A_2, \ldots, A_h \rangle$ where A_{i+1} is derived from A_i by attaching a tile of type t $(i = 0, 1, 2, \cdots, h - 1)$, and tile type t cannot be attached to A_h, then $A_h = A^*$.*

Proof. Let A^* be the least-sized producible assembly such that $A^* \setminus A$ contains only tiles of type t and the frontier of A^* contains no tiles of type t. We will show that A can only grow A^* if only allowed to attach tile type t.

Towards a contradiction, assume there exists a sequence of assemblies from A such that $A_h \neq A^*$. If A_h is some subassembly of A^*, note that we may still attach tiles of type t to reach A^*, implying that A_h does not fit the specified requirements. Otherwise, let A_n be the first assembly in the sequence which contains a tile not in A^*. Consider A_{n-1}. There is no tile of type t attachable to

A_{n-1} such that the tile is not in A^*. If there were, that tile of type t would be attachable to A^*, contradicting the definition of A^*. Therefore no such A_n can exist, implying that A_h must be A^*.

Theorem 8. *There does not exist a $\mathcal{O}(1)$ space unstable concentrations robust fair coin flip tile system in the aTAM.*

Proof. Towards a contradiction, assume that a space-n solution does exist.

As the assembly process proceeds, the key point to consider is when the current assembly enters a state in which multiple distinct positions may attach a tile. In such a case select one type t of all attachable tiles, and increase its concentration to ensure, with high probability, that assembly proceeds by attaching only tiles of type t up until there is no position to attach type t tiles. Such a type t is called a *dominate* type. Let the concentration of the dominate tile type t be $(1 - \frac{1}{100n^2})$. For each step i, let t_i denote the dominate type of concentration $(1 - \frac{1}{100n^2})$.

When there is more than one position to attach the same type of tile t, we are assured by Lemma 1 that a unique assembly will result after repeatedly placing tiles of type t (in any order) until placement of t is no longer an option.

Given this setup, we have that at each step i, the assembly does not grow with a dominate type with probability at most $\frac{1}{10n^2}$. With probability at most $\frac{1}{10n}$, there is a step i among n steps that the assembly does not grow with the dominate type.

Therefore, there is a terminal assembly that will be generated with probability at least 0.9. This is a contradiction.   □

Motivated by the impossibility of robust coin flipping in the aTAM under unstable concentrations, we now consider some established extensions of the aTAM from the literature. In particular, we show that robust coin flipping with unstable concentrations is possible within the aTAM with negative glues [10,18,22], the polyTAM [13], the hexTAM [7] with negative glues, and the GTAM [14].

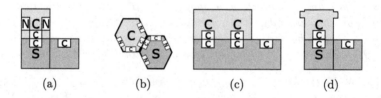

(a)   (b)   (c)   (d)

Fig. 7. The terminal assemblies representing "heads" in some alternate models. C is a strength-τ glue and N is a strength-(-1) glue in (a) the aTAM tile system and (b) the hexTAM tile system. (c) C is a strength-1 glue in a $\tau = 2$ polyTAM tile system. (d) C is a strength-1 glue in a $\tau = 1$ GTAM tile system. The abutting geometry does not allow two C tiles to attach.

Theorem 9. *There exists a $\mathcal{O}(1)$ space unstable concentration robust fair coin-flip tile system in the aTAM with negative glues, polyTAM, hexTAM with negative glues, and the GTAM.*

Proof. Consider a tile assembly system $\Gamma = (T, \sigma, \tau)$ with 3 producible assemblies: σ, a terminal assembly *heads*, and a terminal assembly *tails*. Further, $\sigma \to_1^\Gamma heads$ and $\sigma \to_1^\Gamma tails$. Let $t_{\sigma \to heads}$ and $t_{\sigma \to tails}$ be the same tile c, then $TRANS(\sigma, heads) = TRANS(\sigma, tails) = \frac{P(c)}{2P(c)} = \frac{1}{2}$. Systems which meet these characteristics within the mentioned models can be seen in Fig. 7.   □

6   Conclusions and Future Work

In this paper we have introduced the problem of designing robust, fair coin flipping systems. Generating such coin flips is fundamental for the implementation of randomized self-assembly algorithms. By incorporating concentration independent robustness into the design of such systems, we directly address the practical issue of limited control over species concentrations. Our goal in this work is to provide a stepping stone for the creation of general, robust randomized self-assembly systems. As evidence towards the feasibility of this goal, we have shown how our gadgets can be applied to convert a large class of linear systems into equivalent systems with the concentration robustness property. A more general open problem is as follows: given a general tile system, is it possible to convert the system to an approximately equivalent system that is concentration robust? If possible, how efficiently can this be accomplished in terms of scale factor and approximation factor?

Another direction for future work is the consideration of generalizations of the coin flip problem. Our partition definition for coin flip systems extends naturally to distributions with more than two outcomes, as well as non-uniform distributions. What general probability distributions can be assembled in $O(1)$ space, and with what efficiency? We have also introduced the online variant of concentration robustness in which species concentrations may change at each step of the self-assembly process. We have shown that when such changes are completely arbitrary, coin flipping is not possible in the aTAM. A relaxed version of this robustness constraint could permit concentration changes to be bounded by some fixed rate. In such a model, how close to a fair flip can a system guarantee in terms of the given rate bound? As an additional relaxation, one could consider the problem in which an initial concentration assignment may be *approximately* set by the system designer, thereby modeling the limited precision an experimenter can obtain with a pipette.

A final line of future work focusses on applying randomization in self-assembly to computing functions. The parallelization within the abstract tile assembly model allows for substantially faster arithmetic than what is possible in non-parallel computational models [16]. Can randomization be applied to solve these problems even faster? Moreover, there are a number of potentially interesting problems that might be helped by randomization, such as primality testing, sorting, or a general simulation of randomized boolean circuits.

References

1. Becker, F., Rapaport, I., Rémila, É.: Self-assemblying classes of shapes with a minimum number of tiles, and in optimal time. In: Arun-Kumar, S., Garg, N. (eds.) FSTTCS 2006. LNCS, vol. 4337, pp. 45–56. Springer, Heidelberg (2006)
2. Bryans, N., Chiniforooshan, E., Doty, D., Kari, L., Seki, S.: The power of nondeterminism in self-assembly. In: Proceedings of the 22nd ACM-SIAM Symposium on Discrete Algorithms, SODA 2011, SIAM, pp. 590–602 (2011)
3. Chandran, H., Gopalkrishnan, N., Reif, J.: Tile complexity of linear assemblies. SIAM J. Comput. **41**(4), 1051–1073 (2012)
4. Chen, H.-L., Goel, A.: Error free self-assembly using error prone tiles. In: Ferretti, C., Mauri, G., Zandron, C. (eds.) DNA 2004. LNCS, vol. 3384, pp. 62–75. Springer, Heidelberg (2005)
5. Cheng, Q., Aggarwal, G., Goldwasser, M.H., Kao, M.-Y., Schweller, R.T., de Espanés, P.M.: Complexities for generalized models of self-assembly. SIAM J. Comput. **34**, 1493–1515 (2005)
6. Cook, M., Fu, Y., Schweller, R.T.: Temperature 1 self-assembly: deterministic assembly in 3D and probabilistic assembly in 2D. In: Proceedings of the 22nd ACM-SIAM Symposium on Discrete Algorithms, SODA 2011, pp. 570–589 (2011)
7. Demaine, E.D., Demaine, M.L., Fekete, S.P., Patitz, M.J., Schweller, R.T., Winslow, A., Woods, D.: One tile to rule them all: simulating any tile assembly system with a single universal tile. In: Esparza, J., Fraigniaud, P., Husfeldt, T., Koutsoupias, E. (eds.) ICALP 2014. LNCS, vol. 8572, pp. 368–379. Springer, Heidelberg (2014)
8. Doty, D.: Theory of algorithmic self-assembly. Commun. ACM **55**(12), 78–88 (2012)
9. Doty, D.: Randomized self-assembly for exact shapes. SIAM J. Comput. **39**(8), 3521–3552 (2010)
10. Doty, D., Kari, L., Masson, B.: Negative interactions in irreversible self-assembly. In: Sakakibara, Y., Mi, Y. (eds.) DNA 16 2010. LNCS, vol. 6518, pp. 37–48. Springer, Heidelberg (2011)
11. Doty, D., Lutz, J.H., Patitz, M.J., Summers, S.M., Woods, D.: Random number selection in self-assembly. In: Calude, C.S., Costa, J.F., Dershowitz, N., Freire, E., Rozenberg, G. (eds.) UC 2009. LNCS, vol. 5715, pp. 143–157. Springer, Heidelberg (2009)
12. Evans, C.: Crystals that count! physical principles and experimental investigations of dna tile self-assembly. Ph.D. thesis, California Institute of Technology (2014)
13. Fekete, S.P., Hendricks, J., Patitz, M.J., Rogers, T.A., Schweller, R.T.: Universal computation with arbitrary polyomino tiles in non-cooperative self-assembly. In: Proceedings of the 25th ACM-SIAM Symposium on Discrete Algorithms, SODA 2015, SIAM, pp. 148–167 (2015)
14. Fu, B., Patitz, M.J., Schweller, R.T., Sheline, R.: Self-assembly with geometric tiles. In: Czumaj, A., Mehlhorn, K., Pitts, A., Wattenhofer, R. (eds.) ICALP 2012, Part I. LNCS, vol. 7391, pp. 714–725. Springer, Heidelberg (2012)
15. Kao, M.-Y., Schweller, R.T.: Randomized self-assembly for approximate shapes. In: Aceto, L., Damgård, I., Goldberg, L.A., Halldórsson, M.M., Ingólfsdóttir, A., Walukiewicz, I. (eds.) ICALP 2008, Part I. LNCS, vol. 5125, pp. 370–384. Springer, Heidelberg (2008)
16. Keenan, A., Schweller, R., Sherman, M., Zhong, X.: Fast arithmetic in algorithmic self-assembly. In: Ibarra, O.H., Kari, L., Kopecki, S. (eds.) UCNC 2014. LNCS, vol. 8553, pp. 242–253. Springer, Heidelberg (2014)

17. Patitz, M.J.: An introduction to tile-based self-assembly and a survey of recent results. Natural Comput. **13**(2), 195–224 (2014)
18. Patitz, M.J., Schweller, R.T., Summers, S.M.: Exact shapes and turing universality at temperature 1 with a single negative glue. In: Cardelli, L., Shih, W. (eds.) DNA 17 2011. LNCS, vol. 6937, pp. 175–189. Springer, Heidelberg (2011)
19. Rothemund, P.W.K., Winfree, E.: The program-size complexity of self-assembled squares (extended abstract). In: Proceedings of the 32nd ACM Symposium on Theory of Computing, STOC 2000, pp. 459–468 (2000)
20. Soloveichik, D., Winfree, E.: Complexity of self-assembled shapes. SIAM J. Comput. **36**(6), 1544–1569 (2007)
21. von Neumann, J.: Various techniques used in connection with random digits. J. Res. Natl Bur. Stan. **12**, 36–38 (1951)
22. Winfree, E.: Algorithmic self-assembly of DNA. Ph.D. thesis, California Institute of Technology (1998)

New Geometric Algorithms
for Fully Connected Staged Self-Assembly

Erik D. Demaine[1], Sándor P. Fekete[2(✉)],
Christian Scheffer[2], and Arne Schmidt[2]

[1] CSAIL, MIT, Cambridge, MA, USA
edemaine@mit.edu
[2] Department of Computer Science, TU Braunschweig, Braunschweig, Germany
{s.fekete,c.scheffer,arne.schmidt}@tu-bs.de

Abstract. We consider *staged self-assembly systems*, in which square-shaped tiles can be added to bins in several stages. Within these bins, the tiles may connect to each other, depending on the *glue types* of their edges. Previous work by Demaine et al. showed that a relatively small number of tile types suffices to produce arbitrary shapes in this model. However, these constructions were only based on a spanning tree of the geometric shape, so they did not produce full connectivity of the underlying grid graph in the case of shapes with holes; designing fully connected assemblies with a polylogarithmic number of stages was left as a major open problem. We resolve this challenge by presenting new systems for staged assembly that produce fully connected polyominoes in $\mathcal{O}(\log^2 n)$ stages, for various scale factors and temperature $\tau = 2$ as well as $\tau = 1$. Our constructions work even for shapes with holes and uses only a constant number of glues and tiles. Moreover, the underlying approach is more geometric in nature, implying that it promised to be more feasible for shapes with compact geometric description.

1 Introduction

In *self-assembly*, a set of simple *tiles* form complex structures without any active or deliberate handling of individual components. Instead, the overall construction is governed by a simple set of rules, which describe how mixing the tiles leads to bonding between them and eventually a geometric shape.

The classic theoretical model for self-assembly is the *abstract tile-assembly model* (aTAM). It was first introduced by Winfree [12,14]. The *tiles* used in this model are building blocks , which are unrotatable squares with a specific glue on each side. Equal glues have a connection strength and may stick together. The *glue complexity* is the number of different glues, while the *tile complexity* is the number of tile types. If an additional tile wants to attach to the existing assembly by making use of matching glues, the sum of corresponding glue strengths needs to be at least some minimum value τ, which is called the *temperature*.

A generalization of the aTAM called the *two-handed assembly model* (2HAM) was introduced by Demaine et al. [4]. While in the aTAM, only individual tiles

© Springer International Publishing Switzerland 2015
A. Phillips and P. Yin (Eds.): DNA 2015, LNCS 9211, pp. 104–116, 2015.
DOI: 10.1007/978-3-319-21999-8_7

can be attached to an existing intermediate assembly, the 2HAM allows attaching other partial assemblies. If two partial assemblies ("supertiles") want to assemble, then the sum of the glue strength along the whole common boundary needs to be at least τ.

In this paper we consider the *staged tile assembly model* introduced in [4], which is based on the 2HAM. In this model the assembly process is split into sequential stages that are kept in separate bins, with supertiles from earlier stages mixed together consecutively to gain some new supertiles. We can either add a new tile to an existing bin, or we pour one bin into another bin, such that the content of both get mixed. Hence, there are bins at each stage. Unassembled parts get removed. The overall number of necessary stages and bins are the *stage complexity* and the *bin complexity*. Demaine et al. [4] achieved several results summarized in Table 1. Most notably, they presented a system (based on a spanning tree) that can produce arbitrary polyomino shapes P in $\mathcal{O}(diameter)$ many stages, $\mathcal{O}(\log N) = \mathcal{O}(\log n)$ bins and a constant number of glues, where N is the number of tiles of P, n is the size of a smallest square containing P, and the diameter is measured by a shortest path within P, so it can be as big as N. The downside is that the resulting shapes are not fully connected. For achieving full connectivity, only the special case of monotone shapes was resolved by a system with $\mathcal{O}(\log n)$ stages; for hole-free shapes, they were able to give a system with full connectivity, scale factor 2, but $\mathcal{O}(n)$ stages. This left a major open problem: designing a staged assembly system with full connectivity, polylogarithmic stage complexity and constant scale factor for general shapes.

Our Results. We show that for any polyomino, even with holes, there is a staged assembly system with the following properties, both for $\tau = 2$ and $\tau = 1$.

1. polylogarithmic stage complexity,
2. constant glue and tile complexity,
3. constant scale factor,
4. full connectivity.

See Table 1 for an overview. The main novelty of our method is to focus on the underlying geometry of a constructed shape P, instead of just its connectivity graph. This results in bin numbers that are a function of k, the number of vertices of P: while k can be as big as $\Theta(n^2)$, n can be arbitrarily large for fixed k, implying that our approach promises to be more suitable for constructing natural shapes with a clear geometric structure.

Related Work. As mentioned above, our work is based on the 2HAM. There is a variety of other models, e.g., see [2]. A variation of the staged 2HAM is the *Staged Replication Assembly Model* by Abel et al. [1], which aims at reproducing supertiles by using *enzyme self assembly*. Another variant is the *Signal Tile Assembly Model* introduced by Padilla et al. [9].

Other related geometric work by Cannon et al. [3] and Demaine et al. [5] considers reductions between different systems, often based on geometric properties. Fu et al. [7] use geometric tiles in a generalized tile assembly model to

Table 1. Overview of results from [4] and this paper. The number of tiles of P is denoted by $N \in \mathcal{O}(n^2)$, n is the side length of a smallest bounding square, while k is the number of vertices of the polyomino, with $k \in \Omega(1)$ and $k \in \mathcal{O}(N)$.

Lines and Squares	Glues	Tiles	Bins	Stages	τ	Scale	Conn.	Planar
Line [4]	3	6	7	$\mathcal{O}(\log n)$	1	1	full	yes
Square — Jigsaw techn. [4]	9	$\mathcal{O}(1)$	$\mathcal{O}(1)$	$\mathcal{O}(\log n)$	1	1	full	yes
Square — $\tau = 2$ (Sect. 2.1)	4	$\mathcal{O}(1)$	$\mathcal{O}(1)$	$\mathcal{O}(\log n)$	2	1	full	yes

Arbitrary Shapes	Glues	Tiles	Bins	Stages	τ	Scale	Conn.	Planar
Spanning Tree Method [4]	2	16	$\mathcal{O}(\log n)$	$\mathcal{O}(diameter)$	1	1	partial	no
Monotone Shapes [4]	9	$\mathcal{O}(1)$	$\mathcal{O}(n)$	$\mathcal{O}(\log n)$	1	1	full	yes
Hole-Free Shapes [4]	8	$\mathcal{O}(1)$	$\mathcal{O}(N)$	$\mathcal{O}(N)$	1	2	full	no
Shape with holes (Sect. 2.2)	7	$\mathcal{O}(1)$	$\mathcal{O}(k)$	$\mathcal{O}(\log^2 n)$	2	3	full	no
Hole-Free Shapes (Sect. 2.2)	7	$\mathcal{O}(1)$	$\mathcal{O}(k)$	$\mathcal{O}(\log n)$	2	3	full	no
Hole-Free Shapes (Sect. 3.1)	18	$\mathcal{O}(1)$	$\mathcal{O}(k)$	$\mathcal{O}(\log^2 n)$	1	4	full	no
Shape with holes (Sect. 3.2)	20	$\mathcal{O}(1)$	$\mathcal{O}(k)$	$\mathcal{O}(\log^2 n)$	1	6	full	no

assemble shapes. Fekete et al. [6] study the power of using more complicated polyominoes as tiles.

Using stages has also received attention in DNA self assembly. Reif [11] uses a stepwise model for parallel computing. Park et al. [10] consider assembly techniques with hierarchies to assemble DNA lattices. Somei et al. [13] use a stepwise assembly of DNA tiles. Padilla et al. [8] include active signaling and glue activation in the aTAM to control hierarchical assembly of Robinson patterns. None of these works considers complexity aspects.

2   Fully-Connected Constructions for $\tau = 2$

In the following, we consider fully connected assemblies for temperature $\tau = 2$. We start by an approach for squares (Sect. 2.1). In Sect. 2.2 we describe how to extend this basic idea to assembling general polyominoes.

2.1 $n \times n$ Squares, $\tau = 2$

For $\tau = 2$ assembly systems, it is possible to develop more efficient ways for constructing a square. The construction is based on an idea by Rothemund and Winfree [12], which we adapt to staged assembly. Basically, it consists of connecting two strips by a corner tile, before filling up this frame; see Fig. 1.

Theorem 1. *There exists a $\tau = 2$ assembly system for a fully connected $n \times n$ square with $\mathcal{O}(\log n)$ stages, 4 glues, 14 tiles and 7 bins.*

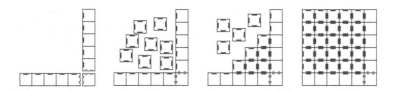

Fig. 1. Construction of fully connected square using $\tau = 2$ and a frame.

Proof. The construction is an easy result of combining known construction for lines by staged assembly with filling in squares in the aTAM with temperature $\tau = 2$, as follows: First we construct the $1 \times (n-1)$ strips with strength-2 glues. We know from [4] that a strip can be constructed in $\mathcal{O}(\log n)$ stages, three glues, six tiles and seven bins. Because both strips are perpendicular, they will not connect. Therefore, we can use all seven bins to construct both strips in parallel. For each strip we use tiles such that the edge toward the interior of the square has a strength-1 glue. In the next stage we mix the single corner tile with the two strips. Finally, we add a tile type with strength-1 glues on all sides. When the square is filled, no further tile can still connect, as $\tau = 2$.

Overall, we need $\mathcal{O}(\log n)$ stages with four glues (three for the construction, one for filling up the square), 14 tiles (six for each of the two strips, one for the corner tile, one for filling up the square) and seven bins for the parallel construction of the two strips. $\qquad\square$

2.2 Polygons with or Without Holes, $\tau = 2$

Our method for assembling a polyomino P at $\tau = 2$ generalizes the approach for building a square that is described in Sect. 2.1. The key idea is to scale P by a factor of 3, yielding $3P$; for this we first build a frame called the *backbone*, which is a spanning tree based on the union of all boundaries of $3P$. This backbone is then filled up in a final stage by applying a more complex version of the flooding approach of Theorem 1. In particular, there is not only one flooding tile, but a constant set S of such distinct tiles.

Definition and Construction of the Backbone. In the following, we consider a scaled copy $3P$ of a polyomino P, constructed by replacing each tile by a 3×3 square of tiles. We define the *backbone* of $3P$ as follows; see Fig. 2 for an illustration.

Definition 1. *A tile of $3P$ is a boundary tile of $3P$, if one of the tiles in its eight (axis-parallel or diagonal) neighbor tiles does not belong to $3P$. A boundary strip of $3P$ is a maximal set of boundary tiles that forms a contiguous (vertical or horizontal) strip, see Fig. 2(b). A boundary component C is a connected component of boundary tiles; because of the scaling, an inside boundary component corresponds to precisely one inside boundary of $3P$ (delimiting a hole), while the outside boundary component corresponds to the exterior boundary of $3P$, see Fig. 2(c). Furthermore, each boundary component C has a unique decomposition*

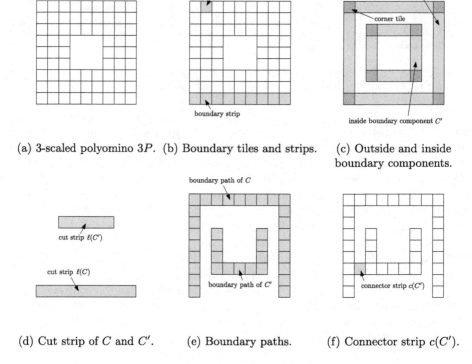

(a) 3-scaled polyomino 3P. (b) Boundary tiles and strips.   (c) Outside and inside boundary components.

(d) Cut strip of C and C'.   (e) Boundary paths.   (f) Connector strip c(C').

Fig. 2. Stepwise construction of the backbone of a 3-scaled polyomino $3P$.

into boundary strips: a circular sequence of boundary strips that alternate between vertical and horizontal, with consecutive strips sharing a single ("corner") tile. For an inside boundary component C', its cut strip $\ell(C')$ *is the leftmost of its topmost strips; for the outside boundary component, its cut strip $\ell(C)$ is the leftmost of its bottommost strips, see Fig. 2(d).*

A boundary path *of the outside boundary component C consists of the union of all its strips, with the exception of $\ell(C)$; for an inside boundary component C', it consists of $C' \setminus \ell(C')$; see Fig. 2(e). Furthermore, the* connector strip $c(C')$ *for an inside boundary component C' is the contiguous horizontal set of tiles of $3P$ extending to the left from the leftmost bottommost tile of C' and ending with the first encountered other boundary tile of $3P$; see Fig. 2(f). Then the* backbone *of $3P$ is the union of all boundary paths and the connector strips of inside boundary components.*

By construction, the backbone has a canonical decomposition into boundary strips and connector strips; furthermore, a tile in the backbone of $3P$ is part of three different strips if and only if it is an end tile of a connector strip. For h holes, only $2h$ tiles in the backbone are part of three different strips.

Overall, this yields a hole-free shape that can be constructed efficiently.

Lemma 1. *Let k be the number of vertices of a 3-scaled polyomino $3P$. The corresponding backbone can be assembled in $\mathcal{O}(\log^2 n)$ stages with 4 glues, $\mathcal{O}(1)$ tiles and $\mathcal{O}(k)$ bins.*

Proof. The main idea is to give a recursive separation of the backbone into trivial tiles such that its reversed order implies a staged self-assembly that fulfills the required guarantees. For separating the backbone, we observe that it consists of two types of components: strips and corner tiles (see Fig. 3). The degree of such a corner tile is two or three, corresponding to the number of adjacent strips.

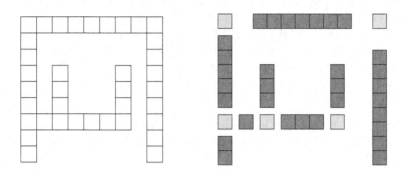

Fig. 3. (Left) A backbone of a polyomino. (Right) A backbone decomposed into strips (green) and corner tiles (yellow) (Color figure online).

The corner tiles will be the splitting points of the separation. The separation of the backbone can be described by three steps. In a first step, we decompose the polyomino by recursively removing the corner tiles of degree three, until only tiles with degree two are left in all components. In the second step, these components are further decomposed via the corner tiles of degree two, such that only strips remain. In a third and final step, the straight strips are decomposed, until just trivial tiles are left.

Because each corner tile has either two or three adjacent strips, the recursive separations of the backbone have a degree of at most three. The key ingredient for an efficient separation, i.e., a polylogarithmic recursion depth, for all three steps, is that the splitting points are chosen such that the sizes of the split components are balanced. In particular, for each splitting, we ensure that the size of each split component is at most the half of the size of the original component. This can be obtained by picking the respective tree median of a remaining backbone piece, i.e., at a corner whose removal leaves each connected component with at most half the number of strips of the original tree. Performing this splitting operation recursively yields a recursion depth of at most $\mathcal{O}(\log k)$.

Proof details can be found in the full version of the paper.

For the overall backbone assembly, we use four glues, $\mathcal{O}(1)$ tiles and $\mathcal{O}(k)$ bins within $\mathcal{O}(\log n \cdot \log k)$ stages: we split at tree medians $\mathcal{O}(\log k)$ times, and use $\mathcal{O}(\log n)$ stages for each strip. $\qquad\square$

By applying the approach of the backbone, we construct any polyomino by assembling its backbone and then flooding it by the set of tiles S, which is illustrated in Fig. 4. To guarantee that flooding the backbone does not exceed the original boundaries, we apply the following property of temperature-2 assemblies.

Lemma 2. *Consider an arbitrary supertile P' and an arbitrary set of single tiles S', such that all non-bonded glues of P' and S' have strength 1. Each position p that is bounded (indirectly) by two parallel, non-glued sides of P' will not be part of any supertile that can be assembled by P' and S' at $\tau = 2$.*

Proof. Consider a position p that is bounded by two parallel, non-glued sides of P' and that is adjacent to two sides of P'. Then at most one of the two sides of P' can have glues. If there is a tile in S' that wants to attach there, the connection strength cannot exceed 1. Hence, the tile cannot be attached to P'. This trivially holds for positions that are adjacent to one side of P' only and for positions that are not adjacent to any side of P'. Thus, Lemma 2 holds.   □

On the one hand, Lemma 2 ensures that no position outside of $3P$ is filled by flooding the backbone. On the other hand, similar as in Theorem 1, the following simple observation guarantees that all tiles of $3P$ that do not belong to the backbone are filled by a single tile:

Property 1. In the configuration of Lemma 2, p is filled by a tile in a unique self-assembly, if and only if there are sequences of collinear, adjacent tiles t_1 and t_2 that fulfill the following:

- t_1 and t_2 are constructible from S' and P',
- t_1 and t_2 meet in p, and
- t_1 and t_2 start from perpendicular unglued sides f_1 and f_2 of P' that bound p (indirectly).

The flooding tiles of S are defined such that their combination with the not bonded glues of the backbone meets the properties of Lemma 2 and Property 1. Hence, by mixing the backbone and S in a single bin leads finally to a fully connected version of $3P$.

Theorem 2. *Let P be an arbitrary polyomino with k vertices. Then there is a $\tau = 2$ staged assembly system that constructs a fully connected version of P in $\mathcal{O}(\log^2 n)$ stages, with 7 glues, $\mathcal{O}(1)$ tiles, $\mathcal{O}(k)$ bins and scale factor 3.*

Proof. For the staged self-assembly of P, we still need to give a set of flooding tiles S and have to define how the sides of the backbone have to be marked by glues such that the flooding leads to a fully connected version of P.

By Lemma 2 and Property 1, we know that every position that does not belong to the polyomino needs to be bounded by at least two parallel unglued sides and every tile of the polyomino must be bounded by at least two perpendicular glued sides. We can construct the backbone while satisfying these properties as follows: we cover each 3-scaled tile of P according to the glue chart, illustrated

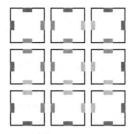

Fig. 4. Glue chart for 3×3 tiles for filling up the shape. Blue glue $\overset{\wedge}{=} g_1$, orange glue $\overset{\wedge}{=} g_2$ and red glue $\overset{\wedge}{=} g_3$ (Color figure online).

in Fig. 4 and mark each side of the backbone's boundary, except for the polyomino's and holes' boundaries, by the glue that is induced by the glue chart (see Fig. 5, middle).

An example for a correct placements of the glues can be found in Fig. 5. Observe that there exist some strips that have a glue type on their end that is different from the glue type that is used in the strip construction (see the red circles in Fig. 5). For building those strips we have to modify the backbone assembly. We assemble those strips completely, but without the last tile with the different glue. Then we add a single tile such that the glue is on the correct place.

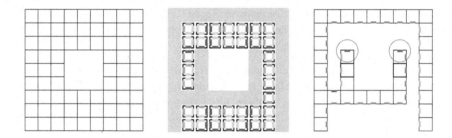

Fig. 5. (Left) A scaled polyomino with one hole. (Middle) Construction of the glues inside the backbone. (Right) Every tile of the backbone is furnished with glues. A strip may have conflicts with glues while the strip is assembling (red circle) (Color figure online).

Overall, we have four glue types for building the backbone, and three glue types for the five strip types and the connection tiles if the side points to the interior of the polyomino. Hence, we use a total of seven glue types and $\mathcal{O}(1)$ tile types.

To fill up the polyomino, we mix the nine kinds of tiles (see Fig. 4) plus the backbone in one bin. In total, we need $\mathcal{O}(\log^2 n)$ stages, seven glues, $\mathcal{O}(1)$ tiles and $\mathcal{O}(k)$ bins to assemble a fully connected polyomino, scaled by a factor 3 from the target shape.   □

As noted before, the number of degree-3 corner tiles depends on the number of holes. We can describe the overall complexity in terms of h, the number of holes. For the special case of hole-free shapes, we can skip some steps, reducing the necessary number of stages. In particular, Corollary 1 follows from Theorem 2.

Corollary 1. *The stage complexity of Theorem 2 can be quantified in the the number of holes h such we get a stage complexity of $O(\log^2 h + \log n)$. In particular, Theorem 2 gives a staged self-assembly system for hole-free shapes with $O(\log n)$ stages, seven glues, $O(1)$ tiles, $O(k)$ bins and a scale factor of 3.*

3   Fully-Connected Constructions for $\tau = 1$

In this section we describe approaches for assembling polyominoes at temperature $\tau = 1$.

3.1   Hole-Free Shapes, $\tau = 1$

We present a system for building hole-free polyominoes. The main idea is based on [4], i.e., splitting the polyomino into strips. Each of these strips gets assembled piece by piece; if there is a component that can attach to the current strip, we create it and attach it.

Our geometric approach partitions the polyomino into rectangles and uses them to assemble the whole polyomino. Even for complicated shapes with many vertices, this number of rectangles is never worse than quadratic in the size of the bounding box; in any case we get a large improvement in the stage complexity.

We first consider a building block, see Fig. 6.

Lemma 3. *A $2n \times 2m$ rectangle (with $n \geq m$) with at most two tabs at top and left side and at most two pockets at each bottom or right side (see Fig. 6) can be assembled with $\mathcal{O}(\log n)$ stages, 9 glues, $\mathcal{O}(1)$ tiles and $\mathcal{O}(1)$ bins at $\tau = 1$.*

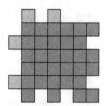

Fig. 6. A square (green) with tabs on top and left side (orange) and pockets on bottom and right side (Color figure online).

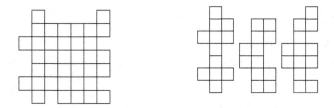

Fig. 7. (Left) A modified square with tabs and pockets. (Right) A partition into components.

Proof. First consider the $2n \times 2m$ square, which we partition into (vertical) rectangles of width 2. As shown in Fig. 7, these are joined by tabs and pockets in rows n and $n + 1$. The glues on their sides are the same as for recursively cutting a square according to the jigsaw technique of [4]. Now every component has a maximum width of 3, even with the tabs. This allows us to use nine glues to create each component with attached tabs and pockets as follows:

A component without a tab or pocket is cut between the $(n - 1)$st and the nth row, as well as between the $(n + 1)$st and the $(n + 2)$nd row. Then we have two strips of width 2 and one 2×2 square. The square can be assembled by brute force with desired glues on its sides. The strips can also be decomposed recursively like a $1 \times n$ strip with desired glues on the sides. Thus, for this kind of component nine glues suffice and the component is built within $\mathcal{O}(\log n)$ stages. Note that we need $\mathcal{O}(1)$ bins to store every possible component of this kind, i.e., they use one out of three possible glue triples on each side.

A component with tabs and/or pockets is cut between rows, such that only components without tabs and pockets and at most four components with tabs and pockets exists. Note that the four components are either single tiles or 1×2 strips. The other components are either strips of width two or similar to the components above. Hence, we need at most $\mathcal{O}(\log n)$ stages to build the biggest component. Then we assemble all components by successively putting together pairs. We observe that this kind of component appears at most six times. Thus, we need six bins to store the components of this kind. Again the nine glues suffice.

Now we have all components in $\mathcal{O}(1)$ bins, so we can assemble the components in a pairwise fashion to the desired polyomino within $\mathcal{O}(\log n)$ stages. Overall our nine glues suffice, so we have $\mathcal{O}(1)$ tiles.   □

Theorem 3. *Let P be a hole-free polyomino with k vertices. Then there is a $\tau = 1$ staged assembly system that constructs a fully connected version of P in $\mathcal{O}(\log^2 n)$ stages, with 18 glues, $\mathcal{O}(1)$ tiles, $\mathcal{O}(k)$ bins and scale factor 4.*

Proof. We cut the polyomino with horizontal lines, such that all cuts go through reflex vertices, leaving a set of rectangles. If V_r is the set of reflex vertices in the polyomino, we have at most $|V_r| =: k_r$ cuts and therefore $\mathcal{O}(k)$ rectangles. Now we find one rectangle that forms a tree median in the rectangle adjacency tree (i.e., a rectangle that splits the tree into connected components that have at most half the number of rectangles). Recursing over this splitting operation, we get a tree decomposition of depth $\mathcal{O}(\log k)$. On the pieces, we use a scale factor of 2 for employing a jigsaw decomposition.

When removing a rectangle R, the remaining polyomino may be split into a number of components; see Fig. 8. To connect all of them to R, we employ another scale factor of 2, allowing us to split R in half with a horizontal line. Each half is further subdivided vertically into jigsaw components, such that each component can connect to a part of R independently from the others, as shown in the figure. When all components have been attached to some part of R, we can assemble both halves of R and then assemble these two together. Doing this for all rectangles produces $\mathcal{O}(k_r)$ new components. Hence, our decomposition tree has at most $\mathcal{O}(k_r) = \mathcal{O}(k)$ leafs, where the leafs are rectangles that need $\mathcal{O}(\log n)$ for construction. This yields $\mathcal{O}(\log k \log n)$ stages overall; the rectangle components consume $\mathcal{O}(k)$ bins. Similar to assembling a square, we need nine glues to uniquely assemble all rectangles to the correct polyomino.

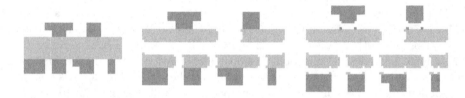

Fig. 8. (Left) A chosen rectangle (orange) that splits the polyomino into components (green). (Middle) Decomposition of splitting rectangle. (Right) Decomposition of the components (Color figure online).

By construction, every rectangle component has at most four adjacent rectangle components; its size is $2w \times 2h$ for some width w and height h. The four adjacent components are all connected at different sides, so the left and upper side each have two tabs, while the right and lower side have two pockets. Thus, we can use the approach of Theorem 3 to assemble all rectangles with 9 additional glues and $\mathcal{O}(1)$ bins for each rectangle component.

Overall, we have $\mathcal{O}(\log n)$ stages to assemble the $\mathcal{O}(k)$ rectangles with $\mathcal{O}(1)$ bins for each rectangle, plus $\mathcal{O}(\log^2 n)$ stages to assemble the polyomino from the rectangles, for a total of $\mathcal{O}(\log^2 n)$ stages and $\mathcal{O}(k)$ bins. For the rectangles we need nine glues, along with nine glues for the remaining assembly, for a total of 18 glues, with $\mathcal{O}(1)$ tile types. The overall scale factor is 4.   □

3.2   Polygons with Holes, $\tau = 1$

Theorem 4. *Let P be an arbitrary polyomino with k vertices. Then there is a $\tau = 1$ staged assembly system that constructs a fully connected version of P in $\mathcal{O}(\log^2 n)$ stages, with 20 glues, $\mathcal{O}(1)$ tiles, $\mathcal{O}(n)$ bins and scale factor 6.*

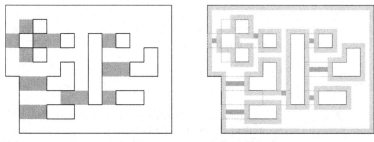

(a) Connections between boundaries.   (b) Tunnels inside the connections.

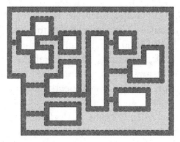

(c) Final shapes S_1 (dark) and S_2 (light).

Fig. 9. Separation of the polyomino P into the two shapes S_1 and S_2.

Proof. (Sketch) Complete details can be found in the full version of the paper; see Fig. 9 for the overall construction. From a high-level point of view, the approach constructs two supertiles S_1 and S_2 separately and finally glues them together, see Fig. 9 for the overall construction. The first supertile S_1 consists of the boundaries of all holes, the boundary of the whole polyomino, and connections between these boundaries. The second supertile S_2 is composed of the rest of the polyomino. A scale factor of 6 guarantees that S_2 is hole-free, which in turns allows employing the approach of Theorem 3. ☐

4   Future Work

Our new methods have the same stage and bin complexity as previous work about stage assemblies [4] and use just a small number of glues. Because the bin complexity is in $\mathcal{O}(k)$ for a polyomino with k vertices, we may need many bins if the polyomino has many vertices. Hence, all our methods are excellent for shapes with a compact geometric description. This still leaves the interesting challenge of designing a staged assembly system with similar stage, glue and tile complexity, but a better bin complexity for polyominoes with many vertices, e.g., for $k \in \Omega(n^2)$?

Another interesting challenge is to develop a more efficient system for an arbitrary polyomino. Is there a staged assembly system of stage complexity $o(\log^2 n)$ without increasing the other complexities?

References

1. Abel, Z., Benbernou, N., Damian, M., Demaine, E.D., Demaine, M.L., Flatland, R., Kominers, S.D., Schweller, R.: Shape replication through self-assembly and rnase enzymes. In: ACM-SIAM Symposium on Discrete Algorithms (SODA), pp. 1045–1064 (2010)
2. Aggarwal, G., Cheng, Q., Goldwasser, M.H., Kao, M.-Y., de Espanes, P.M., Schweller, R.T.: Complexities for generalized models of self-assembly. SIAM J. Comput. **34**(6), 1493–1515 (2005)
3. Cannon, S., Demaine, E.D., Demaine, M.L., Eisenstat, S., Patitz, M.J., Schweller, R.T., Summers, S.M., Winslow, A.:. Two hands are better than one (up to constant factors). In: Symposium on Theoretical Aspects of Computer Science (STACS), pp. 172–184 (2013)
4. Demaine, E.D., Demaine, M.L., Fekete, S.P., Ishaque, M., Rafalin, E., Schweller, R.T., Souvaine, D.L.: Staged self-assembly: nanomanufacture of arbitrary shapes with $o(1)$ glues. Natural Comput. **7**(3), 347–370 (2008)
5. Demaine, E.D., Demaine, M.L., Fekete, S.P., Patitz, M.J., Schweller, R.T., Winslow, A., Woods, D.: One tile to rule them all: simulating any tile assembly system with a single universal tile. In: Esparza, J., Fraigniaud, P., Husfeldt, T., Koutsoupias, E. (eds.) ICALP 2014. LNCS, vol. 8572, pp. 368–379. Springer, Heidelberg (2014)
6. Fekete, S.P., Hendricks, J., Patitz, M.J., Rogers, T.A., Schweller, R.T.: Universal computation with arbitrary polyomino tiles in non-cooperative self-assembly. In: ACM-SIAM Symposium on Discrete Algorithms (SODA), pp. 148–167 (2015)
7. Fu, B., Patitz, M.J., Schweller, R.T., Sheline, R.: Self-assembly with geometric tiles. In: Czumaj, A., Mehlhorn, K., Pitts, A., Wattenhofer, R. (eds.) ICALP 2012, Part I. LNCS, vol. 7391, pp. 714–725. Springer, Heidelberg (2012)
8. Padilla, J.E., Liu, W., Seeman, N.C.: Hierarchical self assembly of patterns from the robinson tilings: DNA tile design in an enhanced tile assembly model. Natural Comput. **11**(2), 323–338 (2012)
9. Padilla, J.E., Patitz, M.J., Schweller, R.T., Seeman, N.C., Summers, S.M., Zhong, X.: Asynchronous signal passing for tile self-assembly: fuel efficient computation and efficient assembly of shapes. Int. J. Found. Comput. Sci. **25**(4), 459–488 (2014)
10. Park, S.H., Pistol, C., Ahn, S.J., Reif, J.H., Lebeck, A.R., Dwyer, C., LaBean, T.H.: Finite-size, fully addressable DNA tile lattices formed by hierarchical assembly procedures. Angewandte Chemie **118**(5), 749–753 (2006)
11. Reif, J.H.: Local parallel biomolecular computation. DNA-Based Comput. **3**, 217–254 (1999)
12. Rothemund, P.W.K., Winfree, E.: The program-size complexity of self-assembled squares (extended abstract). In: ACM Symposium on Theory of Computing (STOC), pp. 459–468 (2000)
13. Somei, K., Kaneda, S., Fujii, T., Murata, S.: A microfluidic device for DNA tile self-assembly. In: DNA Computing (DNA 11), pp. 325–335 (2006)
14. Winfree, E.: Algorithmic self-assembly of DNA. Ph.D thesis, California Institute of Technology (1998)

Leader Election and Shape Formation with Self-organizing Programmable Matter

Zahra Derakhshandeh[1], Robert Gmyr[2], Thim Strothmann[2]([✉]), Rida Bazzi[1], Andréa W. Richa[1], and Christian Scheideler[2]

[1] Computer Science, CIDSE, Arizona State University, Tempe, USA
{zderakhs,bazzi,aricha}@asu.edu
[2] Department of Computer Science, University of Paderborn, Paderborn, Germany
{gmyr,thim}@mail.upb.de, scheideler@upb.de

Abstract. In this paper we consider programmable matter consisting of simple computational elements, called *particles*, that can establish and release bonds and can actively move in a self-organized way, and we investigate the feasibility of solving fundamental problems relevant for programmable matter. As a model for such self-organizing particle systems, we will use a generalization of the geometric amoebot model first proposed in [21]. Based on the geometric model, we present efficient local-control algorithms for leader election and line formation requiring only particles with constant size memory, and we also discuss the limitations of solving these problems within the general amoebot model.

1 Introduction

A central problem for programmable matter is shape formation, and various solutions have already been found for that problem using different approaches like DNA tiles [34], moteins [14], or nubots [39]. We are studying shape formation using the amoebot model which was first proposed in [21]. In order to determine how decentralized shape formation can be handled, we are particularly interested in the connection between leader election and shape formation. In the leader election problem we are given a set of particles, and the problem is to select one of these as the leader. Many problems like the consensus problem (all particles have to agree on some output value) can easily be solved once the leader election problem can be solved. The same has also been observed for shape formation, as most shape formation algorithms depend on some seed element. However, the question is whether shape formation can even be solved in circumstances where leader election is not possible. The aim of this paper is to shed some light on the dependency between leader election and shape formation by focusing on the special problem of forming a line of particles. Before we present our results, we first give a formal definition of the model and the problems we intend to study.

Z. Derakhshandeh and A.W. Richa—Supported in part by the NSF under Awards CCF-1353089 and CCF-1422603.

R. Gmyr, T. Strothmann and C. Scheideler—Supported in part by DFG grant SCHE 1592/3-1.

© Springer International Publishing Switzerland 2015
A. Phillips and P. Yin (Eds.): DNA 2015, LNCS 9211, pp. 117–132, 2015.
DOI: 10.1007/978-3-319-21999-8_8

1.1   Models

We use two models throughout this work. Firstly, we consider a generalization of the amoebot model [21] which abstracts from any geometry information. We call this model the *general amoebot model*. Secondly, we consider a model that is essentially equivalent to the original amoebot model presented in [21] but is defined based on the general amoebot model. We refer to this second model as the *geometric amoebot model*.

In the *general amoebot model*, programmable matter consists of a uniform set of simple computational units called particles that can move and bond to other particles and use these bonds to exchange information. The particles act asynchronously and they achieve locomotion by expanding and contracting, which resembles the behavior of amoeba.

As a base of this model, we assume that we have a set of particles that aim at maintaining a connected structure at all times. This is needed to prevent the particles from drifting apart in an uncontrolled manner like in fluids and because in our case particles communicate only via bonds. The shape and positions of the bonds of the particles mandate that they can only assume discrete positions in the particle structure. This justifies the use of a possibly infinite, undirected graph $G = (V, E)$, where V represents all possible positions of a particle (relative to the other particles in their structure) and E represents all possible transitions between positions.

Each particle occupies either a single node or a pair of adjacent nodes in G, i.e., it can be in two different *shapes*, and every node can be occupied by at most one particle. Two particles occupying adjacent nodes are *connected*, and we refer to such particles as *neighbors*. Particles are *anonymous* but the bonds of each particle have unique labels, which implies that a particle can uniquely identify each of its outgoing edges. Each particle has a local memory, and any pair of connected particles has a shared memory that can be read and written by both particles.

Particles move through *expansions* and *contractions*: If a particle occupies one node (i.e., it is *contracted*), it can expand to an unoccupied adjacent node to occupy two nodes. If a particle occupies two nodes (i.e., it is *expanded*), it can contract to one of these nodes to occupy only a single node. Performing movements via expansions and contractions has various advantages. For example, it would easily allow a particle to abort a movement if its movement is in conflict with other movements. A particle always knows whether it is contracted or expanded and this information will be available to neighboring particles. In a *handover*, two scenarios are possible: a) a contracted particle p can "push" a neighboring expanded particle q and expand into the neighboring node previously occupied by q, forcing q to contract, or b) an expanded particle p can "pull" a neighboring contracted particle q to a cell occupied by it thereby expanding that particle to that cell, which allows p to contract to its other cell. The ability to use a handover allows the system to stay connected while particles move (e.g., for particles moving in a worm-like fashion). Note that while expansions and contractions may represent the way particles physically move in space, they can also be interpreted as a particle "looking ahead" and establishing new logical connections (by expanding) before it fully moves to a new position and severs the old connections it had (by contracting).

Summing up over all assumptions above, the *state* of a particle is uniquely determined by its shape, the contents of its local memory, the edges it has to neighboring particles, the contents of their shared memory (which may allow a particle to obtain further information about the neighboring particles beyond their shape), and finally the shape of the neighboring particles. The *state of the particle system* (or short, *system state*) is defined as the combination of all particle states. We say a particle system in a system state in which the particle occupy a set of nodes $A \subseteq V$ is *connected* if the graph $G|_A$ induced by A is connected. We assume the standard asynchronous computation model, i.e., only one particle can be active at a time. Whenever a particle is active, it can perform an *action* (governed by some fixed, finite size program controlling it) consisting of a finite amount of computation (involving its local memory, the shared memories with its neighboring particles, and random bits) followed by no or a single movement. Hence, a *computation* of a particle system is a potentially infinite sequence of actions $A_1, A_2, \ldots$ based on some initial system state s_0, where action A_i transforms system state s_{i-1} into system state s_i. The (parallel) time complexity of a computation is usually measured in *rounds*, where a round is over once every particle has been given the chance to perform at least one action.

Let $\mathcal{S}$ be the set of all system states in which the particle system is connected. In general, a *computational problem* P for the particle system is specified by a set $\mathcal{S}' \subseteq \mathcal{S}$ of permitted initial system states and a mapping $F : \mathcal{S}' \to 2^{\mathcal{S}}$, where $F(s) \subseteq \mathcal{S}$ determines the set of permitted *final* states for any initial state $s \in \mathcal{S}'$. A particle system *solves* problem $P = (\mathcal{S}', F)$ if for any initial system state $s \in \mathcal{S}'$, all computations of the particle system eventually reach a system state in $F(s)$ without losing connectivity, and whenever such a system state is reached for the first time, the system stays in $F(s)$. If for all computation a final state is reached in which all particles decided to halt (i.e., they decided not to perform any further actions, irrespective of future events), then the particle system is also said to *decide* problem P. Note that being in a final state does not necessarily mean that all particles decided to halt. If $\mathcal{S}' = \mathcal{S}$, so *any* initial state is permitted (including arbitrary faulty states, as long as the particle system is connected), then a particle system solving P is also said to be *self-stabilizing*. It is well-known that in general a distributed system solving a problem P cannot decide it and also be self-stabilizing because if so, it would often be possible to come up with an initial state s where a member of the system decides to halt prematurely, disallowing the system to eventually reach a state in $F(s)$.

Besides the general amoebot model, we will also consider the *geometric amoebot model*. The geometric amoebot model is a specific variant of the general amoebot model in which the underlying graph G is defined to be the equilateral triangular graph G_{eqt} (see Fig. 1), and the bonds of the particles are labeled in a consecutive way in clockwise orientation around a particle so that every particle

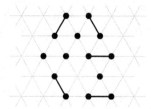 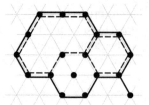

Fig. 1. The left part shows an example of a particle structure in the geometric amoebot model. A contracted particle is depicted as a black dot, and an expanded particle is depicted as two black dots connected by an edge. The right part shows a particle structure with 3 borders. The outer border is shown as a solid line and the two inner borders are shown as dashed lines.

has the same sense of clockwise orientation. However, we do not assume that the labeling is uniform, so the particles do not necessarily share a common sense of direction in the grid.

1.2   Problems

In this paper we consider the following two problems. For both problems we define the set of initial system states as the set of all states such that the particle system is connected and all memories are empty.

For the *leader election problem* the set of final system states contains any state in which the particles form a connected structure and exactly one particle is a leader (i.e., only this particle is in a leader state while the remaining particles are in a non-leader state). Our goal will be to come up with a distributed algorithm that allows a particle system to decide the leader election problem. Note that the leader election problem is well defined for both the general amoebot model and the geometric amoebot model.

In a *shape formation problem*, the set of final states consists of those system states where the particle structure forms the desired shape. As a specific example of a shape formation problem, we consider the *line formation problem*. In the geometric amoebot model, the shape the particles have to form is a straight line in the equilateral triangular grid and all particles have to be contracted in a final system state. Of course, in the general amoebot model a straight line is not well-defined. Hence, for this model the set of final states for the line formation problem is defined to consist of all system states in which the particles form a simple path in G.

Throughout the paper, we assume for the sake of simplicity that in an initial state all particles are contracted. Our algorithms can easily be extended to dispose of this assumption.

1.3   Our Contributions

In this paper we focus on the problem of solving leader election and shape formation for particles with *constant memory*. For shape formation, we just focus on the already mentioned line formation problem.

For the general amoebot model, we can show that neither leader election nor shape formation can be decided by any distributed algorithm. Suppose that there is a distributed algorithm solving the line formation problem in the general amoebot model (when starting in a well-initialized state). Since in this case it is possible to decide when $G|_{A'}$ forms a line, it is also possible to design a protocol that solves the leader election problem: once the line has been formed, its two endpoints contend for leadership using tokens with random bits sent back and forth until one of them wins. On the other hand, one can deduce from [28] that in the general amoebot model there is no distributed algorithm that can decide when a leader has been elected (with any reasonable success probability). More concretely, in [28] the authors show that for the ring of anonymous nodes there is no algorithm that can correctly decide the leader election problem (or in their words, that can solve the leader election problem with distributive termination) with any probability $\alpha > 0$, i.e., for any algorithm in which the particles are guaranteed to halt, the error probability is unbounded. Since in the general amoebot model G can be any graph, we can set G to be a ring whose size is the number of particles and the result of [28] is directly applicable. Hence, there cannot be a distributed algorithm deciding the line formation problem (with any reasonable success probability) in the general amoebot model, and therefore not even an algorithm for solving it since a protocol solving the problem could easily be transformed into a protocol deciding it. However, for the *geometric* amoebot model we show that there is a distributed algorithm that can decide the leader election problem, i.e., at the end we have exactly one leader and the leader knows that it is the only leader left. Moreover, the runtime for our leader election algorithm is worst-case optimal in a sense that it needs at most $O(L)$ rounds on expectation, where L is the maximum length of a border between the particle structure and an empty region (inside or outside of it) in G_{eqt}. Based on the leader election algorithm, we present a distributed algorithm that solves the line formation problem. Both algorithms assume that the system is in a well-initialized state. It would certainly be desirable to have algorithms that can tolerate any initial state, but at the end of the paper we show that there are certain limitations to solving leader election and line formation in a self-stabilizing fashion.

1.4   Related Work

Many approaches related to programmable matter have recently been proposed. One can distinguish between active and passive systems. In passive systems the particles either do not have any intelligence at all (but just move and bond based on their structural properties or due to chemical interactions with the environment), or they have limited computational capabilities but cannot control their movements. Examples of research on *passive systems* are DNA computing [1,8,14,20,37], tile self-assembly systems in general (e.g., see the surveys in [22,34,38]), population protocols [3], and slime molds [9,32]. We will not describe these models in detail as they are only of little relevance for our approach. On the other hand in *active systems*, computational particles can control the way they act and move in order to solve a specific task. Robotic swarms, and modular robotic systems are some examples of active programmable matter systems.

In the area of *swarm robotics* it is usually assumed that there is a collection of autonomous robots that have limited sensing, often including vision, and communication ranges, and that can freely move in a given area. They follow a variety of goals, for example graph exploration (e.g., [23]), gathering problems (e.g., [2,16]), shape formation problems (e.g., [24,35]), and to understand the global effects of local behavior in natural swarms like social insects, birds, or fish (see e.g., [7,11]). Surveys of recent results in swarm robotics can be found in [30,33]; other samples of representative work can be found in e.g., [4,6,17–19,27,31]. While the analytical techniques developed in the area of swarm robotics and natural swarms are of some relevance for this work, the individual units in those systems have more powerful communication and processing capabilities than in the systems we consider.

The field of *modular self-reconfigurable robotic systems* focuses on intra-robotic aspects such as the design, fabrication, motion planning, and control of autonomous kinematic machines with variable morphology (see e.g., [25,40]). *Metamorphic robots* form a subclass of self-reconfigurable robots that share some of the characteristics of our geometric model [15]. The hardware development in the field of self-reconfigurable robotics has been complemented by a number of algorithmic advances (e.g., [10,35,36]), but so far mechanisms that automatically scale from a few to hundreds or thousands of individual units are still under investigation, and no rigorous theoretical foundation is available yet.

The *nubot* model [12,13,39] by Woods et al. aims at providing the theoretical framework that would allow for a more rigorous algorithmic study of biomolecular-inspired systems, more specifically of self-assembly systems with active molecular components. While bio-molecular inspired systems share many similarities with our self-organizing particle systems, there are many differences that do not allow us to translate the algorithms and other results under the nubot model to our systems — e.g., there is always an arbitrarily large supply of "extra" particles that can be added to the system as needed, and the system allows for an additional (non-local) notion of rigid-body movement.

2   Leader Election in the Geometric Amoebot Model

In this section we show how the leader election problem can be decided in the geometric amoebot model. Our approach organizes the particle system into a set of cycles and executes an algorithm on each cycle independently. For simplicity and ease of presentation we first assume that particles have a global view of the cycle they are part of, that agents act synchronously, and that their local memory is unbounded. However, in the local-control protocol none of these assumptions are needed. In particular, the particles only require a constant amount of memory. In Sect. 2.5 we highlight some of the techniques used in the local-control protocol, which relies heavily on token passing. However, due to space constraints the full local-control protocol cannot be presented in detail.

2.1   Organization into Cycles

Let $A \subseteq V$ be any initial distribution of contracted particles such that $G_{\text{eqt}}|_A$ is connected. Consider the graph $G_{\text{eqt}}|_{V \setminus A}$ induced by the unoccupied nodes in G_{eqt}. We call a connected component of $G_{\text{eqt}}|_{V \setminus A}$ an *empty region*. Let $N(R)$ be the neighborhood of an empty region R in G_{eqt}. Then all nodes in $N(R)$ are occupied and we call the graph $G_{\text{eqt}}|_{N(R)}$ a *border*. Since $G_{\text{eqt}}|_A$ is a connected finite graph, exactly one empty region has infinite size while the remaining empty regions have finite size. We define the border corresponding to the infinite empty region to be the unique *outer border* and refer to a border that corresponds to a finite empty region as an *inner border*, see Fig. 1.

The particles occupying a border can instantly (i.e., without communication) organize themselves into a cycle using only local information: Consider a border corresponding to an empty region R. Let p be a particle occupying a node v of the border. By definition there exists a non-occupied node $w \in R$ that is a adjacent to v in the graph G_{eqt}. The particle p iterates over the neighboring nodes of v in clockwise orientation around v starting at w. Consider the first occupied node it encounters; the particle occupying that node is the successor of p in the cycle corresponding to that border. Analogously, p finds its predecessor in the cycle by traversing the neighborhood of v in counter-clockwise orientation.

Note, that a particle can belong to up to three borders at once. Furthermore, a particle cannot locally decide whether two empty regions it sees (i.e., maximal connected components of non-occupied nodes in the neighborhood of v) are distinct. We circumvent these problems by letting a particle treat each empty region in its local view as distinct. For each such empty region, a particle executes an independent instance of the same algorithm. Hence, we say a particle acts as a number of distinct *agents*. For each of its agents a particle determines the predecessor and successor as described above. This effectively connects the set of all agents into disjoint cycles as depicted in Fig. 2. Observe that from a global perspective the cycle of the outer border is oriented clockwise while a cycle of an inner border is oriented counter-clockwise. This is a direct consequence of the way the predecessors and successors of an agent are defined.

2.2   Algorithm

The leader election algorithm operates independently on each cycle. At any given time, some subset of agents on a cycle will consider themselves *candidates*, i.e. potential leaders of the system. Initially, every agent considers itself a candidate. Between any two candidates on a cycle there is a (possibly empty) sequence of non-candidate agents. We call such a sequence a *segment*. For a candidate c we refer to the segment coming after c in the direction of the cycle as $seg(c)$ and refer to its length by $|seg(c)|$. We refer to the candidate coming after c as the *succeeding candidate* ($succ(c)$) and to the candidate coming before c as the *preceding candidate* ($pred(c)$) (see Fig. 3). We drop the c in parentheses if it is clear from the context. We define the distance $d(c_1, c_2)$ between candidates c_1 and c_2 as the number of agents between c_1 and c_2 when going from c_1 to c_2

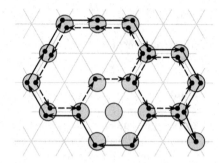

Fig. 2. The depicted particle system is the same as in the right part of Fig. 1. In this figure particles are depicted as gray circles. The black dots inside of a particle represent its agents. As in Fig. 1 the outer border is solid and the two inner borders are dashed.

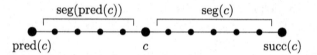

Fig. 3. A cutout from a cycle that is oriented to the right. Non-candidate agents are small black dots, candidates are bigger dots. The candidate c covers $pred(c)$ since $|seg(c)| > d(pred(c), c)$.

in *direction* of the cycle. We say a candidate c_1 *covers* a candidate c_2 (or c_2 *is covered by* c_1) if $|seg(c_1)| > d(c_2, c_1)$ (see Fig. 3). The leader election progresses in *phases*. In each phase, each candidate executes Algorithm 1. A phase consists of three synchronized *subphases*, i.e., agents can only progress to the next subphase once all agents have finished the current subphase.

Consider the execution of Algorithm 1 by a candidate c. If the algorithm returns "not leader" then c revokes its candidacy and becomes part of a segment. If the algorithm returns "leader", c will become the leader of the particle system. The transferal of candidacy in subphase 2 means that c withdraws its own candidacy but at the same time promotes the agent at position *pos* (i.e., $succ(c)$ in subphase 1) to be a candidate. Once a candidate becomes a leader it broadcasts this information such that all particles can halt.

2.3   Correctness

In order to show the correctness of our algorithm, we show that it satisfies the following conditions, that relate to the entire particle system (not just a single cycle):

1. *Safety*: There always exists at least one candidate.
2. *Liveness*: In each phase if there is more than one candidate, at least one candidate withdraws leadership with a probability that is bounded below by a positive constant.

Algorithm 1. Leader Election for a candidate c

Subphase 1:
$pos \leftarrow$ position of $succ(c)$
if covered by any candidate or $|seg(c)| < |seg(pred(c))|$ **then**
    return not leader

Subphase 2:
if coin flip results in heads **then**
    transfer candidacy to agent at pos

Subphase 3:
if only candidate on border **then**
    if outside border **then**
        return leader
    else
        return not leader

Lemma 1. *Algorithm 1 satisfies the safety condition.*

Proof. We will show by induction that on the cycle associated with the outer border there will always be at least one candidate. Initially, this holds trivially. So assume that it holds before a phase. Let c be the candidate with the longest segment. Then there is no candidate covering c and also $|seg(c)| < |seg(pred(c))|$ cannot be true. Hence, c will not withdraw candidacy in subphase 1. In subphase 2, the candidacy of c might be transferred but will not vanish. Let c' be the agent that received the candidacy if it was transferred and $c' = c$ otherwise. In subphase 3, c' will not withdraw candidacy because it lies on the outer border. Hence, there is still a candidate after the phase. □

Lemma 2. *Algorithm 1 satisfies the liveness condition.*

Proof. Assume that there are two or more candidates in the system. First we consider the case that there is a cycle with two or more candidates. If there are segments of different lengths on that cycle, we have $|seg| < |seg(pred)|$ for at least one candidate which will therefore withdraw its candidacy in subphase 1. If all segments are of equal length, we have that in subphase 2 with probability at least $\frac{1}{4}$ there is a candidate c that transfers candidacy while $succ(c)$ does not. Hence, the number of candidates is reduced with probability at least $\frac{1}{4}$. Now consider the case that all cycles have at most one candidate. Then there is a cycle corresponding to an inner border that has exactly one candidate. That candidate will withdraw candidacy in subphase 3 and thereby reduce the number of candidates in the system. □

The following Theorem is a direct consequence of Lemmas 1 and 2.

Theorem 1. *Algorithm 1 decides the leader election problem in the geometric amoebot model.*

2.4 Runtime Analysis

For a cycle of agents let L be the length of the cycle and let l_i be the longest segment length before phase i of the execution of Algorithm 1. We define $l_i = L$ if there is no candidate on the cycle. It is easy to see that if $l_i \geq L/2$ then in phase $i+1$ either the leader is elected (outer border) or all candidates on the cycle vanish (inner border). For the case $l_i < L/2$, Lemma 3 provides the key insight of our analysis.

Lemma 3. *For any phase i such that $l_i < L/2$ it holds $l_{i+1} \geq l_i$ in any case and $l_{i+1} \geq 2l_i$ with probability at least $1/4$.*

Let $L_{\max}$ be the length of the longest cycle in the particle system. Based on Lemma 3 it is easy to see that under complete synchronization of subphases and with the agents having a global view of the cycle, our algorithm requires on expectation $O(\log(L_{\max}))$ phases to elect a leader. For now assume that our algorithm can be realized as a local-control protocol such that phase i requires $O(l_i)$ rounds. Theorem 2 gives a bound on the number of rounds required by the algorithm based on this assumption. The theorem also also holds for the local-control protocol given the definition of a round from Sect. 1.1.

Theorem 2. *Algorithm 1 requires $O(L_{max})$ rounds on expectation.*

Proof. Let the random variable X_i describe the number of rounds during the execution of Algorithm 1 such that $l_i \in [2^{i-1}, 2^i)$. Then, under the assumption that phase i requires $O(l_i)$ rounds, the total runtime of our algorithm is

$$T = \sum_{i=1}^{\lceil \log(L_{\max}) \rceil} X_i \cdot O(2^i).$$

Since $E(X_i) \leq 4$ due to Lemma 3, the expected runtime is

$$E(T) \leq \sum_{i=1}^{\lceil \log(L_{\max}) \rceil} E(X_i) \cdot O(2^i) = O(L_{\max}).$$

$\square$

Note that subphase 1 of the algorithm is not important in terms of correctness. However, it is crucial to achieve a linear runtime in expectation. If agents would only execute subphases 2 and 3, the runtime would degrade to $O(L_{\max} \log L_{\max})$.

2.5 Asynchronous Local-Control Protocol

Here we present some details on how specific parts of the algorithm can be realized as an asynchronous local-control protocol. We focus on the realization of solitude verification and the inner outer border test of subphase 3.

Solitude Verification. A candidate that wants to determine whether it is the only candidate left, tests if its segment ends in another candidate or in itself. To do so, it enforces its own orientation on all agents in its segment. Thereby, every agent in the segment is able to determine the direction of its outgoing edge in direction of the cycle. These edge directions can be seen as vectors in the two dimensional plane and in case the segment is the whole cycle, the vectors cancel out component wise (see Fig. 4). By a simple token passing scheme agents will try match their edge directions component wise with an edge direction in the opposing direction from another agent. Finally, the candidate inspects the segment and if all agents are matched it is the only candidate left on the cycle.

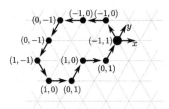

Fig. 4. An example of solitude verification: the candidate (shown slighly bigger) has enforced its orientation (x and y arrows) on all agents. All non-candidate agents determined the offset to the succeeding agent in direction of the cycle (arrows and numbers at nodes).

Inner Outer Border Test. The last candidate of a cycle can decide whether its cycle corresponds to an inner or the outer border as follows. A cycle corresponding to an inner border has counter-clockwise rotation while a cycle corresponding to the outer border has clockwise rotation, see Fig. 2. The candidate sends a token along the cycle that sums the angles of the turns the cycle takes, see Fig. 5. When the token returns to the candidate its value represents the external angle of the polygon corresponding to the cycle while respecting the rotation of the cycle. So it is $-360°$ for an inner border and $360°$ for the outer border. The token can represent the angle as an integer k such that the angle is $k \cdot 60°$. Furthermore, to distinguish the two possible final values of k it is sufficient to store the k modulo 5, so that the token only requires 3 bits of memory.

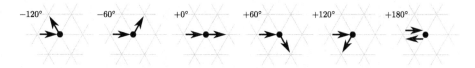

Fig. 5. The angle between the directions a token enters and exits an agent.

3   Line Formation in the Geometric Amoebot Model

Next we consider the line formation problem in the geometric amoebot model. We assume that initially we have an arbitrary connected structure of contracted particles with a unique leader. The leader is used as the starting point for the line of particles and specifies the direction in which this line will grow. As the line grows, every particle touched by the line that is already in a valid line position becomes part of the line. Any other particle connected to the line becomes the root of a tree of particles. Every root aims at traveling around the line in a clockwise manner until it joins the line. As a root particle moves, the other particles in its tree follow in a worm-like fashion (i.e., via a series of handover operations)[1].

Before we give a detailed description of the algorithm, we provide some preliminaries. We distinguish for the state of a particle between *inactive*, *follower*, *root*, and *retired* (or halted). Initially, all particles are *inactive*, except the leader particle, which is always in a *retired* state. In addition to its state, each particle p may maintain a constant number of *flags* in its shared memory. For an expanded particle, we denote the node the particle last expanded into as the *head* of the particle and call the other occupied node its *tail*: In our algorithm, we assume that every time a particle contracts, it contracts out of its tail.

The spanning forest algorithm, given in Algorithm 2, is a basic building block we use for shape formation problems. This algorithm aims at organizing the particles as a spanning forest, where the particles that represent the roots of the trees determine the direction of movement, whom the remaining particles follow. Each particle p continuously runs Algorithm 2 until p becomes retired. If particle p is a follower, it stores a flag $p.parent$ in its shared memory corresponding to the edge adjacent to its parent p' in the spanning forest (any particle q can then easily check if p is a child of q). If p is the leader particle, then it sets the flag $p.linedir$ in the shared memories corresponding to two of its edges in opposite directions (i.e., an edge i and the edge that appears three positions after i in clockwise order), denoting that it would like to extend the line through the directions given by these edges.

We have the following theorem, where *work* is defined as the number of expansions and contractions executed by all particles in the system:

Theorem 3. *Algorithm 2 solves the line formation problem in worst-case optimal $O(n)$ number of rounds and $O(n^2)$ work.*

4   Self-stabilizing Leader Election and Shape Formation

Consider the variant of the geometric amoebot model in which faults can occur that arbitrarily corrupt the local memory of a particle. Recall that for an algorithm to solve the leader election problem in a self-stabilizing manner, it has to

---

[1] For a simulation video of the Line Formation Algorithm please see http://sops.cs.upb.de.

Algorithm 2. Line Formation Algorithm

SPANNINGFOREST (p):

Particle p acts as follows, depending on its current state:

inactive: If p is connected to a retired particle, then p becomes a *root* particle. Otherwise, if an adjacent particle p' is a root or a follower, p sets the flag $p.parent$ on the shared memory corresponding to the edge to p' and becomes a *follower*. If none of the above applies, it remains inactive.

follower: If p is contracted and connected to a retired particle, then p becomes a *root* particle. Otherwise, it considers the following three cases: (i) if p is contracted and p's parent p' (given by the flag $p.parent$) is expanded, then p expands in a handover with p', adjusting $p.parent$ to still point to p' after the handover; (ii) if p is expanded and has a contracted child particle p', then p executes a handover with p'; (iii) if p is expanded, has no children, and p has no inactive neighbor, then p contracts.

root: Particle p may become *retired* following CHECKRETIRE (p). Otherwise, it considers the following three cases: (i) if p is contracted, it tries to expand in the direction given by LINEDIR(p); (ii) If p is expanded and has a child p', then p executes a handover contraction with p'; (iii) if p is expanded and has no children, and no inactive neighbor, then p contracts.

retired: p performs no further action.

CHECKRETIRE (p):

    if p is a contracted root **then**
        if p has an adjacent edge i to p' with a flag $p'.linedir$, where p' is retired **then**
            Let i' be the edge opposite to i in clockwise order
            p sets the flag $p.linedir$ in the shared memory of edges i and i'
            p becomes *retired*.

LINEDIR(p):

    Let i be the label of an edge connected to a retired particle.
    while edge i points to a retired particle **do**
        $i \leftarrow$ label of next edge in clockwise direction
    return i

---

satisfy the following requirements: First, from any initial system state (in which the particle structure is connected) the particle system eventually reaches a final system state while preserving connectivity, i.e., eventually a unique leader will be established. Second, once a final system state is reached, the system has to remain in that state as long as no faults occur. Analogous requirements have to be satisfied for self-stabilizing shape formation.

Our leader election algorithm can be extended to a self-stabilizing leader election algorithm with $O(\log^* n)$ memory using the results of [5,29] (i.e., we use their self-stabilizing reset algorithm on every cycle in order to recover from failure states). However, it is not possible to design a self-stabilizing algorithm for the line formation. The reason for this is that even a much simpler problem called *movement problem* cannot be solved in a self-stabilizing manner. It is easy to see that if the movement problem cannot be solved in a self-stabilizing manner, then also the line formation problem cannot be solved in a self-stabilizing manner.

In the movement problem we are given an initial distribution A of particles that can be in a contracted as well as expanded state, and the goal is to change the set of nodes occupied by the particles without causing disconnectivity. For the ring of expanded particles it holds that for any protocol P there is an initial state so that P does not solve the movement problem. To show this we consider two cases: suppose that there is any state s for some particle in the ring that would cause that particle to contract. In this case set two particles on opposite sides of the ring to that state, and the ring will break due to their contractions. Otherwise, P would not move any particle of the ring, so also in this case it would not solve the movement problem in a self-stabilizing manner.

5   Conclusion

We think that the algorithms presented for the geometric amoebot model can be extended for the case that G is a different regular grid graph embedded in the two-dimensional Euclidean plane. As future work, we would like to identify the minimum set of key geometric properties that G must have in order for the proposed algorithms to work. Also, if in the geometric amoebot model, the particles had a common sense of direction, we would like to investigate whether leader election could be solved deterministically using a slight modification of our algorithm: for each border the last remaining candidate is deterministically chosen to be the "east-most" particle of the set of the "south-most" particles. This algorithm would be similar to the one proposed in [26] for tile self-assembly systems.

References

1. Adleman, L.M.: Molecular computation of solutions to combinatorial problems. Science **266**(11), 1021–1024 (1994)
2. Agathangelou, C., Georgiou, C., Mavronicolas, M.: A distributed algorithm for gathering many fat mobile robots in the plane. In: Proceedings of the 32nd ACM Symposium on Principles of Distributed Computing (PODC), pp. 250–259 (2013)
3. Angluin, D., Aspnes, J., Diamadi, Z., Fischer, M.J., Peralta, R.: Computation in networks of passively mobile finite-state sensors. Distrib. Comput. **18**(4), 235–253 (2006)
4. Arbuckle, D., Requicha, A.: Self-assembly and self-repair of arbitrary shapes by a swarm of reactive robots: algorithms and simulations. Auton. Robots **28**(2), 197–211 (2010)
5. Awerbuch, B., Ostrovsky, R.: Memory-efficient and self-stabilizing network RESET (extended abstract). In: Proceedings of the Thirteenth Annual ACM Symposium on Principles of Distributed Computing (PODC), pp. 254–263 (1994)
6. Barriere, L., Flocchini, P., Mesa-Barrameda, E., Santoro, N.: Uniform scattering of autonomous mobile robots in a grid. Int. J. Found. Comput. Sci. **22**(3), 679–697 (2011)

7. Bhattacharyya, A., Braverman, M., Chazelle, B., Nguyen, H.L.: On the convergence of the hegselmann-krause system. In: Proceedings of the 4th Innovations in Theoretical Computer Science (ITCS), pp. 61–66 (2013)
8. Boneh, D., Dunworth, C., Lipton, R.J., Sgall, J.: On the computational power of DNA. Discrete Appl. Math. **71**, 79–94 (1996)
9. Bonifaci, V., Mehlhorn, K., Varma, G.: Physarum can compute shortest paths. In: Proceedings of the Twenty-Third Annual ACM-SIAM Symposium on Discrete Algorithms (SODA), pp. 233–240 (2012)
10. Butler, Z.J., Kotay, K., Rus, D., Tomita, K.: Generic decentralized control for lattice-based self-reconfigurable robots. Int. J. Robot. Res. **23**(9), 919–937 (2004)
11. B. Chazelle. Natural algorithms. In: Proceedings of the Twentieth Annual ACM-SIAM Symposium on Discrete Algorithms (SODA), pp. 422–431 (2009)
12. Chen, H.-L., Doty, D., Holden, D., Thachuk, C., Woods, D., Yang, C.-T.: Fast algorithmic self-assembly of simple shapes using random agitation. In: Murata, S., Kobayashi, S. (eds.) DNA 2014. LNCS, vol. 8727, pp. 20–36. Springer, Heidelberg (2014)
13. Chen, M., Xin, D., Woods, D.: Parallel computation using active self-assembly. In: Soloveichik, D., Yurke, B. (eds.) DNA 2013. LNCS, vol. 8141, pp. 16–30. Springer, Heidelberg (2013)
14. Cheung, K.C., Demaine, E.D., Bachrach, J.R., Griffith, S.: Programmable assembly with universally foldable strings (moteins). IEEE Trans. Robot. **27**(4), 718–729 (2011)
15. Chirikjian, G.: Kinematics of a metamorphic robotic system. In: Proceedings of the 1994 International Conference on Robotics and Automation (ICRA), pp. 449–455 (1994)
16. Cieliebak, M., Flocchini, P., Prencipe, G., Santoro, N.: Distributed computing by mobile robots: gathering. SIAM J. Comput. **41**(4), 829–879 (2012)
17. Cohen, R., Peleg, D.: Local spreading algorithms for autonomous robot systems. Theor. Comput. Sci. **399**(1–2), 71–82 (2008)
18. Das, S., Flocchini, P., Santoro, N., Yamashita, M.: On the computational power of oblivious robots: forming a series of geometric patterns. In: Proceedings of 29th ACM Symposium on Principles of Distributed Computing (PODC) (2010)
19. Defago, X., Souissi, S.: Non-uniform circle formation algorithm for oblivious mobile robots with convergence toward uniformity. Theor. Comput. Sci. **396**(1–3), 97–112 (2008)
20. Demaine, E.D., Patitz, M.J., Schweller, R.T., Summers, S.M.: Self-assembly of arbitrary shapes using rnase enzymes: meeting the kolmogorov bound with small scale factor (extended abstract). In: Proceedings of the 28th International Symposium on Theoretical Aspects of Computer Science, pp. 201–212 (2011)
21. Derakhshandeh, Z., Dolev, S., Gmyr, R., Richa, A., Scheideler, C., Strothmann, T.: Brief announcement: amoebot – a new model for programmable matter. In: Proceedings of the 26th ACM Symposium on Parallelism in Algorithms and Architectures, (SPAA), pp. 220–222 (2014)
22. Doty, D.: Theory of algorithmic self-assembly. Commun. ACM **55**(12), 78–88 (2012)
23. Flocchini, P., Ilcinkas, D., Pelc, A., Santoro, N.: Computing without communicating: ring exploration by asynchronous oblivious robots. Algorithmica **65**(3), 562–583 (2013)
24. Flocchini, P., Prencipe, G., Santoro, N., Widmayer, P.: Arbitrary pattern formation by asynchronous, anonymous, oblivious robots. Theor. Comput. Sci. **407**(1), 412–447 (2008)

25. Fukuda, T., Nakagawa, S., Kawauchi, Y., Buss, M.: Self organizing robots based on cell structures - cebot. In: Proceedings of the International Conference on Intelligent Robots and Systems, (IROS), pp. 145–150 (1988)

26. Hendricks, J., Patitz, M.J., Rogers, T.A.: Replication of arbitrary hole-free shapes via self-assembly with signal-passing tiles (2015). arXiv preprint arXiv:1503.01244

27. Hsiang, T.-R., Arkin, E., Bender, M., Fekete, S., Mitchell, J.: Algorithms for rapidly dispersing robot swarms in unknown environments. In: Proceedings of the 5th Workshop on Algorithmic Foundations of Robotics (WAFR), pp. 77–94 (2002)

28. Itai, A., Rodeh, M.: Symmetry breaking in distributive networks. In: 22nd Annual Symposium on Foundations of Computer Science, (FOCS), pp. 150–158 (1981)

29. Itkis, G., Levin, L.: Fast and lean self-stabilizing asynchronous protocols. In: 35th Annual Symposium on Foundations of Computer Science, (FOCS), pp. 226–239 (1994)

30. Kernbach, S. (ed.): Handbook of Collective Robotics - Fundamentals and Challanges. Pan Stanford Publishing, Singapore (2012)

31. Kling, P., Meyer auf der Heide, F.: Convergence of local communication chain strategies via linear transformations. In: Proceedings of the 23rd ACM Symposium on Parallelism in Algorithms and Architectures, (SPAA), pp. 159–166 (2011)

32. Li, K., Thomas, K., Torres, C., Rossi, L., Shen, C.-C.: Slime mold inspired path formation protocol for wireless sensor networks. In: Dorigo, M., Birattari, M., Di Caro, G.A., Doursat, R., Engelbrecht, A.P., Floreano, D., Gambardella, L.M., Groß, R., Şahin, E., Sayama, H., Stützle, T. (eds.) ANTS 2010. LNCS, vol. 6234, pp. 299–311. Springer, Heidelberg (2010)

33. McLurkin, J.: Analysis and implementation of distributed algorithms for multi-robot systems. Ph.D. thesis, Massachusetts Institute of Technology (2008)

34. Patitz, M.J.: An introduction to tile-based self-assembly and a survey of recent results. Nat. Comput. 13(2), 195–224 (2014)

35. Rubenstein, M., Cornejo, A., Nagpal, R.: Programmable self-assembly in a thousand-robot swarm. Science 345(6198), 795–799 (2014)

36. Walter, J.E., Welch, J.L., Amato, N.M.: Distributed reconfiguration of metamorphic robot chains. Distrib. Comput. 17(2), 171–189 (2004)

37. Winfree, E., Liu, F., Wenzler, L.A., Seeman, N.C.: Design and self-assembly of two-dimensional dna crystals. Nature 394(6693), 539–544 (1998)

38. Woods, D.: Intrinsic universality and the computational power of self-assembly. In: Neary, T., Cook, M. (eds) Proceedings of the 6th Conference on Machines, Computations and Universality 2013, (MCU), EPTCS, vol. 128, pp. 16–22 (2013)

39. Woods, D., Chen, H., Goodfriend, S., Dabby, N., Winfree, E., Yin, P.: Active self-assembly of algorithmic shapes and patterns in polylogarithmic time. In: Proceedings of the 4th Conference on Innovations in Theoretical Computer Science, ITCS, pp. 353–354 (2013)

40. Yim, M., Shen, W.-M., Salemi, B., Rus, D., Moll, M., Lipson, H., Klavins, E., Chirikjian, G.S.: Modular self-reconfigurable robot systems. IEEE Robot. Autom. Mag. 14(1), 43–52 (2007)

Leakless DNA Strand Displacement Systems

Chris Thachuk[1], Erik Winfree[1], and David Soloveichik[2,3]([✉])

[1] California Institute of Technology, Pasadena, USA
[2] University of California, San Francisco, San Francisco, USA
[3] The University of Texas at Austin, Austin, USA
`david.soloveichik@utexas.edu`

Abstract. While current experimental demonstrations have been limited to small computational tasks, DNA strand displacement systems (DSD systems) [25] hold promise for sophisticated information processing within chemical or biological environments. A DSD system encodes designed reactions that are facilitated by three-way or four-way toehold-mediated strand displacement. However, such systems are capable of spurious displacement events that lead to *leak*: incorrect signal production. We have identified sources of leak pathways in typical existing DSD schemes that rely on toehold sequestration and are susceptible to toe-less strand displacement (i.e. displacement reactions that occur despite the absence of a toehold). Based on this understanding, we propose a simple, domain-level motif for the design of leak-resistant DSD systems. This motif forms the basis of a number of DSD schemes that do not rely on toehold sequestration alone to prevent spurious displacements. Spurious displacements are still possible in our systems, but require multiple, low probability events to occur. Our schemes can implement combinatorial Boolean logic formulas and can be extended to implement arbitrary chemical reaction networks.

1 Introduction

Although biological in origin, nucleic-acids have proven to be versatile materials for *de novo* engineered molecular systems. In particular, cascades of prescribed molecular events can be systematically constructed with so-called strand displacement reactions (DNA strand displacement, abbreviated as DSD). In a basic DSD cascade, one nucleic-acid molecule hybridizes with a partially double-stranded complement, releasing the original binding partner, which in turn triggers a downstream strand displacement event [17]. These interactions can be readily programmed by designing strands to have appropriate complementarity. Dynamic molecular systems like logic circuits [15], amplification schemes [23,26], neural networks [16], as well as mechanical devices like motors [11] have all been experimentally realized [25]. DSD cascades can also emulate any chemical reaction network (CRN, system of chemical reaction equations obeying mass-action rate laws) [1,4,19]. By realizing an appropriate CRN, DSD systems can generate temporal patterns, perform signal processing, remember states, compute distributed algorithms, as well as other tasks that have been studied in the language of CRNs.

© Springer International Publishing Switzerland 2015
A. Phillips and P. Yin (Eds.): DNA 2015, LNCS 9211, pp. 133–153, 2015.
DOI: 10.1007/978-3-319-21999-8_9

However, to move beyond relatively small proof-of-principle demonstrations, we must tackle an important issue — one that has up to now limited the scale of strand displacement systems. DSD systems have been observed to be susceptible to various levels of *leak*: the triggering of undesired strand displacement reactions.

A number of ideas have been proposed to combat leak. For example, designing sequences to have strong C-G bonds at the ends of helixes decreases the rate of fraying, and thus impedes the toeless displacement responsible for leak (see Sect. 2). By further securing helix ends, "clamp" domains of 1 to 3 nucleotides can decrease certain kinds of unintended displacements (see Sect. 3) [15, 17, 23]. Another approach involves adding small quantities of "threshold gates" that preferentially consume leaked strands before they have a chance to interact downstream [15, 17]. The idea is that small leaks get neutralized, but when the desired displacement occurs, the threshold gates are saturated and the signal propagates. Although such thresholding can be effective in the context of digital on/off behavior, it is not fitting for analog or dynamical systems, where information is carried in the temporally varying amount of released signal strands. In particular, the existing leak mitigation options are insufficient for the implementations of CRNs. Not only are such systems strongly analog, but the large concentration differences between "fuel" (aka auxiliary) complexes and signal strands may result in the situation that the amount of leak is comparable to the amount of signal. Other leak reduction strategies include introducing Watson-Crick mismatches [10], and physically segregating different complexes [21].

We are interested in a systematic method capable of reducing leak to arbitrary desired limits. In the form of our argument, we are motivated by "proof-reading" in algorithmic self-assembly, where constructing the same pattern at larger scales (with increasing redundancy) in principle arbitrarily decreases the error rate [2, 22]. Similarly, we describe an ensemble of constructions with different levels of redundancy (parameter N).

The simplest non-trivial DSD operation is sequence translation: a cascade of strand displacement reactions that upon initiation by a strand with sequence X, results in the release of a strand with sequence Y, such that sequences X and Y are unrelated. Note that the output strand can contain some additional "left-over" domains from X as long as these are insufficient to trigger unintended displacement. A pair of translators $Y := W$ and $Y := X$ can be thought as a logical $Y = OR(W, X)$ operation, since either W or X is sufficient to produce Y. In the context of implementing CRNs, translators serve as unimolecular reactions. Thus we start by analyzing regular translators, as well as their "leakproof" counterparts. We later describe AND gates: DSD modules that produce output Y only when both inputs W and X are present. Together with translators, such AND gates are sufficient to implement "dual-rail" Boolean formulas (where abstract signal $X = 1$ is indicated by the presence of active strand $X1$ and absence of active strand $X0$, and vice versa) [17]. At the end of this paper, we discuss the generalization to leakless CRNs.

We argue that with the smallest non-trivial redundancy parameter ($N = 2$), the leakless translator ("double-long domain" (DLD) scheme) exhibits signifi-

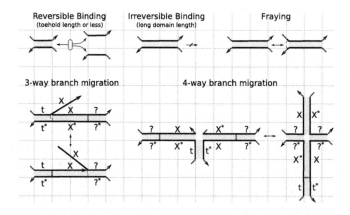

Fig. 1. DNA strand displacement events and conventions.

cantly less leak than a regular translator ("single-long domain" (SLD) scheme). Specifically we show a significant improvement at the usual experimental concentrations (e.g. 100 nM, where the leak is expected to be about five orders of magnitude less), as well as analyze the leak's concentration scaling. If the concentrations decrease by a factor of α, the ratio of the leak of the DLD scheme compared to the SLD scheme is expected to decrease by a factor of α. The rates of the intended reactions are not significantly different for the SLD and DLD translator schemes. To estimate specific leak rates, and to verify that we have not missed any substantial leak pathways, we rely on an automated strand displacement reaction enumerator [9]. The scaling of leak with concentration is obtained analytically.

We then consider more generally translators and AND gates with arbitrary redundancy parameter N. Using a combinatorial argument, we prove that leak requires the joining of N initially separate complexes. We use this to argue that *even at thermodynamic equilibrium*, the amount of activated reporter (i.e. net leak) decreases exponentially with N. Importantly, the thermodynamic argument does not make strong assumptions on the types of reactions possible, or the types of admissible structures. For example, leaks that result in pseudoknots are not excluded. Our proof applies directly to a hypothetical experiment in which a translator or AND gate is composed with a reporter complex (which reports on the output via a fluorescence signal). However, more work is needed to extend this proof to arbitrary composition of leakless gates in networks.

2   Preliminary

We briefly summarize the DNA strand displacement conventions adopted in this paper, which are illustrated in Fig. 1. A detailed overview of these systems, including 3-way and 4-way DNA strand displacement can be found elsewhere [5,25]. We study strand displacement systems at the domain level of abstraction. Domains have a defined length, and are an abstract representation of sequences of that length. We define X^* to be the complement of domain X. Complementary domains can bind while non-complementary domains cannot. Complementary domains bound only by a single grid unit in length, called *toehold domains*, are reversibly bound (i.e., they can spontaneously disassociate). Complementary domains bound by at least two grid units in length, called *long domains*, are considered irreversibly bound (in reality long domains can be arbitrarily longer than the toehold domains – our graphical notation is just for convenience). Valid strand displacement events include ends of bound domains *fraying* (exposing base pairs that can bind to other domains), 3-way branch migration and 4-way branch migration. Binding of complementary domains can lead to two complexes combining into one, while branch migration (possibly followed by toehold unbinding) can lead to one complex separating into two. Unless otherwise qualified, the term *strand displacement* refers to the binding of a toehold domain followed by 3-way branch migration of the neighboring long domain (in other words, toehold mediated strand displacement). We use the term *toeless* strand displacement to refer to 3-way branch migration that is not preceded by binding of toehold-length domains. Rather it mechanistically occurs when double stranded long domains fray, followed by binding of a complementary invading strand to the momentarily opened bases and subsequent branch migration.

For the DNA strand displacement schemes proposed in this paper, we adopt the convention that each long domain X_i can be decomposed into a number of one grid unit parts labelled X_{ia}, X_{ib}, X_{ic}, etc. We use the subscript to denote when a domain is a proper suffix or prefix of X_i. For example, domain X_{iab} consists of the concatenation of X_{ia} and X_{ib}. Note that since X_{iab} is two grid units, it is itself a long domain, and thus irreversibly bound. By standard convention, also followed here, the molecules initially present in a strand displacement system can be classified as either *signal strands* or *fuel complexes*. Signal strands, or signals, propagate information through a system; these are typically at relatively low concentrations. Fuel complexes, typically held at a higher concentration, facilitate reactions by consuming input signals and producing output signals, via a sequence of strand displacement events and intermediate molecules. This process is initiated when input signals displace other strands on a fuel complex. As part of a multistranded complex, we say a signal strand is *sequestered* (or inactive) if it is unable to displace other strands. *External* signal strands carry information from component to component, and *intermediate* signal molecules (strands or complexes) carry information within a single component (e.g. between different fuel molecules). During strand displacement *waste* species can be created and are considered inert.

3 The Single Long Domain (SLD) Motif

Leak in typical DSD systems occurs when the following condition holds: *The unbound part of strand A shares a long domain with the bound part of strand B, but strand A is not currently supposed to displace B.* In the SLD motif, every sequestered strand is bound to a complex by at most one long domain

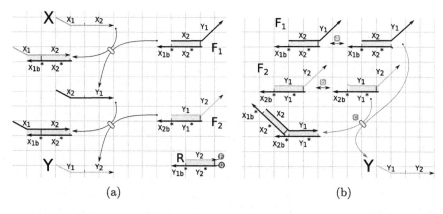

(a)   (b)

Fig. 2. A typical SLD implementation scheme of the translator $Y := X$ that uses *clamp* domains to help combat *leak*. F_1 and F_2 are fuel complexes. The output signal Y is sequestered on F_2 and should only be displaced in the presence of input X. R is a downstream reporter complex that is designed to interact with output signal Y. (a) The intended pathway when X is present consisting of two strand displacement reactions. (b) A leak pathway when X is not present: (reaction a) fraying of the clamp, (reaction b) fraying of the Y_1 long domain, (reaction c) toeless strand displacement via the fleetingly exposed 1-nt toehold at the left of Y_1. When Y is spuriously produced it can successfully interact with downstream reporter complex, R.

and one toehold. Thus when the above condition holds, strand A can toelessly displace strand B. Although this rate is significantly slower than properly toehold mediated strand displacement, it nonetheless occurs at a non-negligible rate for relevant reaction regimes.

Consider a typical translator in a hypothetical experiment in which it is composed with a downstream reporter (Fig. 2a). The reporter emits an experimentally measurable fluorescence signal when the fluorophore (F) is dissociated from the quencher (Q). More generally, the reporter represents downstream complexes which receive input from this translator in a larger circuit. The intended pathway when input X is present is illustrated in Fig. 2a, consisting of two strand displacement reactions.

In the absence of input signal X it is still possible for output signal Y to be produced spuriously. Specifically, fuel F_1 can toelessly displace Y. Similarly, a toeless displacement interaction between F_2 and the reporter R can produce a fluorescence signal in the absence of X. Short *clamp* domains help mitigate these events, but they cannot eliminate them altogether due to fraying of the clamps. One of a number of leak pathways of the translator is illustrated in Fig. 2b.

Enumerator Analysis of the Leak. To systematically analyze the kinetics of the leak pathways we employed an automatic enumerator of strand displacement reactions which can be configured to capture 3-way and 4-way branch migration, toeless and remote toehold displacement, and cooperative hybridization [8,9].

(Since the enumerator ignores interactions forming pseudoknotted structures, and prunes reactions based on a number of assumptions, there could be plausible leak pathways that are not enumerated.) The enumerator input files for systems in this paper are included in the standard release of the enumerator [8]. Using size 5 toeholds, size 15 long domains, and size 2 clamps, the enumerator computes that the net leak rate between F_1 and F_2 resulting in free Y strand is 58 M^{-1} s^{-1} [F_1] $\cdot$ [F_2]. Additionally, there is also a leak between F_2 and reporter R that occurs with rate 13 M^{-1} s^{-1} [F_2] $\cdot$ [R]. For comparison, the intended path in the presence of X consists of three strand displacement reactions which the enumerator predicts occur with rate constants between $4 \cdot 10^5$ M^{-1} s^{-1} and $8 \cdot 10^5$ M^{-1} s^{-1} using the same parameters. The enumerator's rate constant predictions are order-of-magnitude plausible compared to experimentally measured values of 1.4 M^{-1} s^{-1} and $9.6 \cdot 10^5$ M^{-1} s^{-1} for toeless strand displacement and displacement via length-5 toeholds, respectively [27].

4   The Double Long Domain (DLD) Motif

The double long domain (DLD) motif dictates that sequestered signal strands are necessarily bound by at least two consecutive long domains. Schemes that use the DLD motif can be designed to satisfy the following *DLD scheme invariant*: *If a strand A is not intended to displace another strand B then any consecutive, unbound sequence of A differs from the bound sequence of B by at least one long domain (i.e. two grid units in our diagrams).* In other words: there is never sufficient unbound sequence on a single strand to displace another that should not be displaced. This contrasts with typical SLD schemes that rely solely on the absence of open toehold domains to prevent certain reactions.

A DLD translator implementation is shown in Fig. 3a. In the absence of input signal X, it is not possible to fully displace output signal Y (without breaking bonds between long domains). The main leak pathway rather involves domains Y_1 and Y_2 becoming transiently unbound on the same strand due to the interaction of F_1 and F_2. So in the absence of input signal X, reaching a state where the reporter is triggered requires multiple low probability and quickly reversible events to occur, and then not undo before the reporter has a chance to interact.

One possible leak pathway is illustrated in Fig. 3b. First, (reaction a) the clamp of F_2 must fray, then (reaction b) the bound domain Y_1 must fray, (reaction c) fuel F_1 can now form a first base pair, and then (reaction d) proceed to toelessly displace domain Y_1 with 3-way branch migration. At this point consecutive domains Y_1 and Y_2 are open and available to react with the reporter (i.e. signal Y is active). Reactions e and f show two other states in which both domains of the signal Y are active. In reaction e domain X_{2bc} on the signal strand can become bound to the invading complex via "open toehold" 4-way branch migration, which requires initiation by a slow loop-closing event [5]. In reaction f domain X_{2a} induces 3-way branch migration. Importantly, all these reactions are reversible and the states in which Y is active are transient. Then the triggering of the reporter requires catching the fuels in such an active Y state.

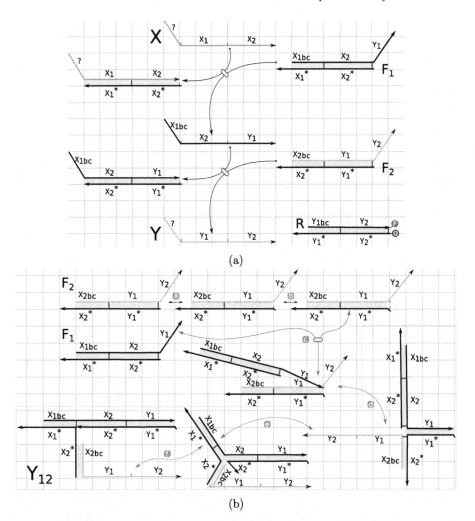

Fig. 3. An implementation of the translator Y := X that uses the DLD motif. (a) The intended pathway when X is present consisting of two strand displacement reactions. R is a downstream reporter complex designed to interact with output signal Y. (b) Low probability leak pathway when input signal X is not present. See text for the description of reactions $a - f$. Transient complexes that have a functional Y signal and can successfully interact with downstream reporter complex, R, are shown outlined in yellow.

Enumerator Analysis of the Leak. The full DLD translator plus reporter system is too large for the current version of the enumerator to analyze at the necessary level of detail for evaluating leaks. As a result we decomposed the enumeration of the leak into two sub-problems: reversibly generating a complex with Y_1 and Y_2 domains open, and then the reaction of this complex with the reporter.

The enumerator reports that the net rate at which F_1 and F_2 react to produce a complex (call it ER) capable of reacting with the reporter is $3 \cdot 10^{-6}$ M^{-1} s^{-1} $\cdot$ $[F_1] \cdot [F_2]$. The reverse reaction occurs with rate constant $4.7 \cdot 10^{-4}$ s^{-1}. The reaction with the reporter is predicted to occur at $5.7 \cdot 10^5$ M^{-1} s^{-1} $\cdot [ER] \cdot [R]$. Since we can upper bound the concentration of complex ER by its equilibrium value, the overall rate at which leak is produced can be estimated to be:

$$3400 M^{-2} s^{-1} \cdot [F_1] \cdot [F_2] \cdot [R].$$

The enumerator also confirms that no interactions between the reporter and either fuel individually can result in the separation of the two reporter strands.

Both the existence (or non-existance) of reaction pathways and their rates will depend on the the the assumptions used by the enumerator.[1] We adjusted enumerator parameters to strike a balance between ensuring that potential leak pathways were explored and yet combinatorial complexity remained tractable – which, on top of the uncertainty in the rate formulas, suggests that enumerator results should be regarded as provisional.

Comparing SLD and DLD Leak Rates. For typical 100 nM concentrations of fuels and reporter, SLD leak rate is roughly five orders of magnitude larger than DLD leak rate. Further, consider how the leak scales with the concentrations of the fuel complexes and the reporter. The SLD leak is a product of two concentrations, while the DLD leak scales as a product of three — the DLD leak effectively acts as a trimolecular reaction. The intended reaction pathway is bimolecular for both SLD and DLD schemes. Since the leak pathway is bimolecular for the SLD scheme, decreasing or increasing concentrations should not change the ratio of leak to intended rates. However, for the DLD scheme, decreasing the concentrations should linearly decrease the ratio of leak to intended rates.

DLD AND Gates. Going beyond translators, the basic computational DSD primitive has historically been the AND gate. We give two distinct AND gate constructions that maintain the DLD scheme invariant. Both are AND gates in the sense that signal X is produced (i.e. domains X_1 and X_2 are unbound) if and only if both input signals A and B are present. They also have the property that if one input is missing, the other input is not permanently consumed. The first scheme is simpler, while the second scheme has the potential advantage of better scaling to low concentrations.

---

[1] These assumptions include the approximate rate formulas for domain-level steps such as hybridization, fraying, 3-way and 4-way branch migration. There are also parameters set by the user that control the potential combinatorial explosion of the enumeration process, such as the granularity of domains (dividing a domain into subdomains allows the enumerator to explore more potential leak pathways, but makes the combinatorics worse), and the relevant time scales (opening a long double-stranded domain is "too slow to consider" and will not be enumerated, while a branch migration pathway may be "too fast" for considering bimolecular interactions prior to the end point).

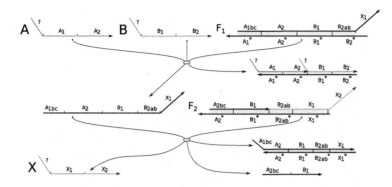

Fig. 4. An implementation of $X := \mathrm{AND}(A, B)$ using the DLD motif and cooperative hybridization.

The scheme of Fig. 4 employs "cooperative" strand displacement [24]. Input signals A and B cooperate to displace the intermediate signal strand from fuel F_1, which in turn displaces the output signal strand X bound to F_2. The overall process is driven forward since the final state contains three additional bound toeholds and maintains the same number of complexes as the initial state. Note, however, that the cooperative step (input signals A and B displacing the intermediate signal from fuel F_1) is effectively trimolecular.[2] Thus, in contrast to the DLD translator, the intended displacement rate for the cooperative DLD AND gate (i.e. when both inputs are present) decreases just as quickly as the leak rate as concentrations are decreased.

The scheme of Fig. 5 employs "associative" strand displacement [3,7,13]. Two consecutive strand displacements (first with input A and then with input B) must occur on fuel complex F_1 to create a displacing complex with open domains A_1 and B_3, capable of displacing the intermediate signal sequestered in F_2. In other words these two strand displacement reactions "glue" (or "associate") domains A_1 and B_3 together. The subsequent interaction with F_3 and F_4 is similar to a DLD translator, except the last step involves opening of a loop. The structure of F_4 is designed to ensure that another invariant is maintained: there is no signal strand that has open domains X_1 and X_2 but not X_3. This invariant ensures that when the second reaction occurs in a downstream complex, it can only occur with the full X_1, X_2, X_3 signal strand, and does not become irreversibly blocked without properly gluing X_3. The overall process is driven forward since the final state contains 4 additional bound toeholds and maintains the same number of complexes as the initial state. Although somewhat more complex, the associative scheme has the advantage that it lacks trimolecular

---

[2] Each partial displacement is reversible and quickly reaches a pseudo-equilibrium proportional to two concentrations (F_1 and an input). The second input then reacts, for an overall rate proportional to the product of $[F_1] \cdot [A] \cdot [B]$.

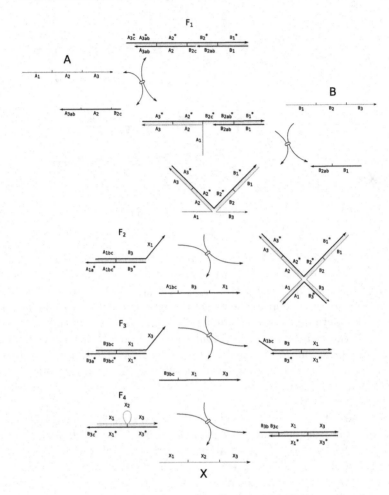

Fig. 5. An implementation of $X := \mathrm{AND}(A, B)$ using the DLD motif and associative hybridization.

steps,[3] and thus the associative AND gate should be faster than the cooperative AND gate in low concentration regimes.

The Triple Long Domain (TLD) Motif. The DLD motif can naturally be generalized to the triple long domain (TLD) motif, where each active signal is represented by three consecutive unbound long domains. The TLD translator is shown in Fig. 6b, and an example leak pathway is shown in Fig. 6c. In

---

[3] Note that although the first reaction is reversible, the reverse reaction is bimolecular as opposed to unimolecular as is the case with the partial displacement by one input in the cooperative AND gate. Thus it is not as readily reversible, especially in low concentration regimes, and the associative gate avoids effectively trimolecular steps.

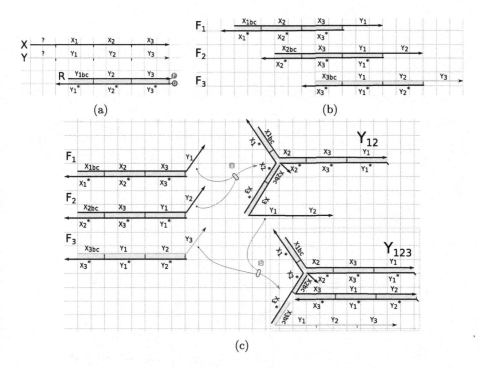

(a)

(b)

(c)

Fig. 6. An implementation of the translator $Y := X$ that uses the triple long domain (TLD) motif. (For simplicity, we show this scheme without clamps.) (a) Input signal and output signal format, and downstream reporter complex, R. (b) Fuel complexes. (c) A coarse grained leak pathway showing only bimolecular reactions.

this pathway, the reporter interacts with a transient structure (called Y_{123}) in which domains Y_1, Y_2 and Y_3 are open. Intuitively, *all three* fuel complexes come together in a transient structure (Y_{123}) — a process which is more unlikely than two fuel complexes coming together in the DLD motif. Indeed, this idea is naturally generalized to signals represented by N long domains as described in the next section.

5   The N Long Domain (NLD) Motif

In this section, with the goal of decreasing the leak *arbitrarily*, we generalize the DLD motif to the NLD motif (*"N-long domains"*, with arbitrary redundancy parameter N). We show constructions for translators and AND gates, and develop an asymptotic thermodynamic argument supporting the claim of leak reduction. In contrast to the main contributions of the previous section,

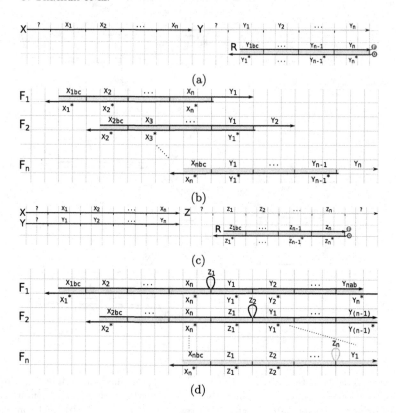

Fig. 7. Reaction implementations that use the N long domain (NLD) motif. (a) Input signal and output signal for translator $Y := X$, and downstream reporter complex, R. (b) Fuel complexes for translator $Y := X$. When input is present, at each step of the intended pathway, the intermediate signals that propagate through the cascade lose one X domain and gain one Y domain. (c) Input signal and output signal for $Z := \mathrm{AND}(X, Y)$, and downstream reporter complex, R. (d) Fuel complexes for $Z := \mathrm{AND}(X, Y)$. When both inputs are present, at each step of the intended pathway, the intermediate signals that propagate through the cascade lose one X domain and one Y domain and gain one Z domain (which is initially in a loop on the fuel complex).

the leak pathway does not necessarily require toeless displacement.[4] Rather, the argument more generally reflects the decreasing likelihood of a sequence of thermodynamically unfavorable reactions. Since the argument is thermodynamic, it makes few assumptions on types of reactions and types of structures possible. The analysis even extends to leaks that involve pseudoknots.

---

[4] For example, in the NLD AND gate described below, if input X is present, then there is a sequence of toehold-mediated reactions that can trigger the reporter. In particular, X displaces the top strand of F_1 from the left up to the hairpin, which in turn displaces the top strand of F_2 from the left up to the hairpin, and so forth. However, each of these reactions would quickly reverse because the partial displacement leaves each top strand attached. The associative hybridization AND gate of Fig. 5 also exhibits this behavior.

Figure 7b and d show the general NLD (i.e. redundancy N) translator and AND gates. The intended operation of the NLD translator when the input is present consists of a cascade of N strand displacement reactions. (The thermodynamic driving force is the formation of N new toehold bonds. The number of separate molecules before and after is the same, and thus there is no effective entropic driving force or penalty.) The intended operation of the AND gate when both inputs are present consists of cooperative strand displacement on F_1, followed by a cascade of $N-1$ toehold mediated strand displacement reactions. (The thermodynamic driving force is the formation of $N+1$ new toehold bonds. While the number of separate molecules decreases by 1 as a result of cooperative strand displacement, this entropic penalty is overcome by the enthalpic gain of the toehold bonds.) With increasing N we expect the kinetics of the desired pathway to slow down quadratically with N.[5]

What about the leak pathway? Unlike for the DLD scheme, we do not examine the kinetics of the pathway. Rather we found we can bound the total amount of leak as we increase N, even if we allow the system to operate indefinitely (and reach thermodynamic equilibrium). In the rest of the section we develop an asymptotic thermodynamic argument examining the enthalpic and entropic driving and opposing forces for the leak in the general NLD scheme, leading to the conclusion that leak decreases exponentially with N.

5.1 Thermodynamics of Leak with Increasing N

The thermodynamic argument in this section relies on two assumptions about toehold and long domains. *(1) The binding of two separate molecules is thermodynamically unfavorable if only one new toehold bond forms ($\Delta G > 0$, bounded away from 0 independently of N). (2) The binding of long domains is effectively irreversible ($\Delta G \ll 0$).* The span of the regime where these assumptions are valid is determined by the concentrations involved, by the temperature, and by the length and sequence composition of the domains. Keeping these parameters fixed, we consider an asymptotic argument on increasing N.

In this section, we say *bottom strand* to refer to the strands that are illustrated on the bottom of their complex in Fig. 7b–d (and analogously *top strand*.) Similarly, we say a domain is a *bottom domain* if it occurs on a bottom strand and is therefore written as starred. Likewise *top domains* occur on top strands and are unstarred.

Imagine hypothetical experiments in which an NDL translator $Y := X$, or AND gate $Z := \text{AND}(X, Y)$, are paired with a reporter reading out their output. In the case of the translator, input X is absent; in the case of the AND gate, either input X or Y (or both) are absent. We do not assume a single molecule

---

[5] The increasing length of the branch migration region is expected to lead to a linearly decreasing success probability per collision [20]. Thus each of the N strand displacement reactions slows down linearly with N. The time spent in the random walk of branch migration will increase quadratically, but will not be rate limiting for practical concentrations and values of N.

experiment; each present molecule can be present in many copies. Leak occurs when the bottom and top strands of a reporter molecule separate. We devote the following section (Sect. 5.2) to proving that producing a single activated reporter molecule necessarily involves the binding of N top fuel strands and N bottom fuel strands in a single complex together with the reporter bottom strand (assuming (2) above). Thus our intuition based on the TLD scheme is confirmed. To understand what this implies about the equilibrium amount of leak let us examine the enthalpic and entropic gains and costs.

Consider the thermodynamic equilibrium of the hypothetical translator and AND gate experiments described above. We argue that (assuming (1) above) the free energy difference between leaked and initial configurations increases linearly with N (with the leaked configurations unfavorable). What is the enthalpic driving force for leak? For the worst case, consider the AND gate when input X is present and input Y is absent: Note that there are N unbound bottom domains $(X_1^*, \ldots, X_N^*)$ that could potentially become bound in some leaked state if input X is present (all other single stranded domains have no complementary domains in the whole system). What is the entropic cost of leak? If any leaked configuration involves the binding of N fuel top strands and N fuel bottom strands in a single complex together with the reporter bottom strand, then the number of separate molecules decreases by N.

Assumption (1) implies that the formation of N new toehold bonds at the cost of decreasing the number of separate complexes by N is thermodynamically unfavorable, with the net increase in free energy proportional to N. This means that the amount of translator or AND gate in the leaked configuration decreases exponentially with N at thermodynamic equilibrium.

Does our thermodynamic argument about the individual translator and AND gate generalize to a circuit of these components? Unfortunately, the combinatorial argument in the next section fails to account for other types of strands with domains that overlap with the given translator or AND gate, as would occur in upstream and downstream components. Also note that in a circuit of these components, a leaked upstream signal may enable non-leak downstream reactions gaining N toehold bonds *per downstream component*. When we consider a fixed circuit, and increase N, this means that leak may result in αN new toehold bond formed, where $\alpha \geq 1$ depends on the circuit. To ensure that the amount of leak at the output of the whole circuit decreases exponentially with N, we need to be in a regime where forming α toehold bonds but losing entropy due to combining two separate molecules is still unfavorable — a strengthening of assumption (1).

5.2   Combinatorics: Leak Requires Binding of All Fuels

Translator. Recall the hypothetical experiment on the NDL translator paired with a downstream reporter (Fig. 7b). Suppose input X is absent, but the fluorophore and quencher are on separate molecules. We now show that assuming we do not decrease the number of bonded long domains (per assumption (2) above), there are N fuel top strands and N fuel bottom strands bound in the same complex as the bottom strand of the leaked reporter.[6]

We ignore toeholds for the rest of the argument and assume for simplicity that the top strands are extended all the way to the left. Let C be the complex containing the bottom strand but not the top strand of the reporter. First we show that C must contain the same number of top fuel strands as bottom fuel strands. Consider top domain X_N and the complementary bottom domain X_N^*. Without the input, there is the same number of domains X_N as X_N^* in the system, and we say that X_N is *balanced*. Note that the initial state has the maximum number of long domain bonds. Thus, if we don't decrease the sum count of long domain bonds, every X_N must be bonded to a X_N^* and vice versa in C as well. Since each top fuel strand contains exactly one X_N domain, and every bottom fuel strand contains exactly one X_N^* domain, it must be that complex C contains the same number of top fuel strands as bottom fuel strands.

Let t and b be the total number of top domains and bottom domains in complex C, respectively. If C contains s top fuel strands (and therefore s bottom fuel strands), then the difference $t - b$ is: $s \cdot (N + 1) - s \cdot (N) - N$. The first term captures the contribution of $N + 1$ top domains in each top fuel strand. The second term captures the contribution of N bottom domains in each bottom fuel strand. The last term captures the contribution of the N bottom domains on the bottom reporter strand.

If we maximize the number of long domain bonds, it must be that complex C contains at least as many top domains as bottom domains (otherwise, we have an unbound bottom domain but we started with all bottom domains bound.) Thus, $s \cdot (N + 1) - s \cdot (N) - N \geq 0$. This implies that $s \geq N$. In other words, complex C contains N top fuel strands and N bottom fuel strands.

And Gate. Recall the hypothetical experiment on the NDL AND gate paired with a downstream reporter (Fig. 7d). Suppose one of the inputs is absent. We show that if the reporter top strand (fluorophore) is not in the same complex as the reporter bottom strand (quencher), then the complex containing the reporter bottom strand contains N fuel complexes (assuming we do not decrease the number of bonded long domains).

Again let C be the complex containing the bottom reporter strand but not the top reporter strand. If either input X or input Y is absent, then either X_N or Y_1 domains are balanced (i.e. have the same number of top as bottom types). Every fuel top strand contains exactly one X_N and Y_1 domain and every fuel

---

[6] Note that it is not enough to notice that each fuel complex has one top Y domain in excess and thus assume that to replace the top reporter strand requires all N fuel complexes. As we saw before, there are possible cascades between fuel complexes that need to be taken into account. To drive home the point, consider removing the leftmost X_1 and X_1^* domains from F_1. Then we could swap the top strands on F_1 and F_2 without decreasing the number of long domains bound, and then F_1 will contain two open Y domains: Y_1 and Y_2. In this case, only $N - 1$ fuel complexes are sufficient to replace the top reporter strand.

bottom strand contains exactly one X_N^* and Y_1^* domain. Thus if one of the inputs is absent, the complex C contains the same number of fuel top strands as fuel bottom strands.

Suppose input X is absent. Suppose C contains s top fuel strands (and therefore s bottom fuel strands). Let t be the total count of top X and Z domains in complex C, and let b be the total count of bottom X and Z domains in complex C. Note that the contribution of any fuel top strand (and respectively fuel bottom strand) to the difference $t - b$ is identical, and thus we can avoid considering exactly *which* fuel strands are in complex C. Since each fuel top strand contains a total of $N + 1$ of top X and Z domains, and each fuel bottom strand contains a total of N of bottom X and Z domains, the difference $t - b$ is: $s \cdot (N+1) - s \cdot (N) - N$. The last term again is the contribution of the N bottom Z domains on bottom reporter strand. Since globally there is an excess of top Z domains (and a balanced amount of X domains), it must be that on complex C, the above difference is non-negative. In other words, $s \cdot (N+1) - s \cdot (N) - N \geq 0$, which implies that $s \geq N$.

If, on the other hand, input Y is absent (and X is possibly present), we proceed as before but let t be the total count of top Y and Z domains in complex C, and let b be the total count of bottom Y and Z domains in complex C. Note that similarly each fuel top strand contains a total of $N + 1$ of top Y and Z domains, and each fuel bottom strand contains a total of N of bottom Y and Z domains. Since without the Y input, there is a balanced amount of Y domains, and as before an excess of top Z domains, we again must satisfy the same inequality $s \cdot (N+1) - s \cdot (N) - N \geq 0$, which implies $s \geq N$.

6   CRNs Using the DLD Motif

Can general chemical reaction networks be emulated using leakless DNA strand displacement systems? As a proof of principle, we give two implementations for the canonical reaction $A + B \rightarrow X + Y$ using the DLD motif. We note that this reaction can emulate any bimolecular reaction with at most two reactants and at most two products — A or B or both can be declared as *fuel* species (whose concentration is assumed to be constant) for reactions with less than two reactant input signals, whereas X or Y or both can be declared as *waste* species (which are considered inert) for reactions with less than two product signals. Furthermore, higher order reactions, or bimolecular reactions with more than two products, can be emulated by a cascade of these canonical reactions.

How can we determine if these schemes are *correct*, even in the absence of leak? Several general approaches to this question have been explored for DSD implementations [6,12,18]. The task is divided into two parts: enumerating all the (non-leak) reactions between the molecules (either using an enumerator [9,14] or formal proofs) followed by applying some notion of program equivalence between the original CRN and the implemented CRN. Unfortunately, none of these approaches is fully satisfactory when applied to the systems described here. For example, Lakin, Phillips, and Stefanovic have previously shown that any

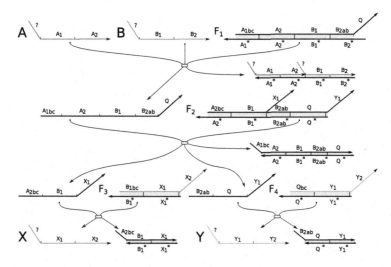

Fig. 8. An implementation of $A + B \rightarrow X + Y$ using the DLD motif and cooperative hybridization. Domain Q is unique to reactions that have $A + B$ as reactants.

DNA strand displacement encoding scheme satisfying certain modularity properties can be formally proven to correctly implement a chemical reaction network of interest, in terms of the notion of serializability [12]. Our proposed schemes do not satisfy the sufficient conditions to claim correctness with respect to serializability: both schemes have been optimized to share intermediates between different reaction encodings, and are therefore not strictly modular. However, both schemes do satisfy a key property identified in the serializability result: each DSD reaction cascade, called a *reaction encoding*, that emulates a formal reaction should be *transactional*. Informally, a reaction encoding $r_1, \ldots, r_i, \ldots, r_n$ that emulates the formal reaction $A + B \rightarrow X + Y$ is transactional if it can be partitioned into two parts, separated by its first irreversible reaction, r_i. The reactions $r_1, \ldots, r_{i-1}$ must be reversible and cannot produce signals representing X nor Y, and the signals representing formal species A and B must be consumed in $r_1, \ldots, r_i$. The reactions $r_i, \ldots, r_n$ must produce the signals representing X and Y. The enumerator tool used in previous sections was applied to both scheme's encodings of $A + B \rightarrow X + Y$, and the resulting implementation CRNs were then easily verified to be transactional in this sense. At this point, we have not yet established that in implementations of larger CRNs, no unexpected crosstalk interactions between molecules would arise during the enumeration.

6.1 DLD Motif + Cooperative Hybridization

The DLD motif can be used in combination with cooperative hybridization [24] to implement arbitrary chemical reaction networks. The implementation of the

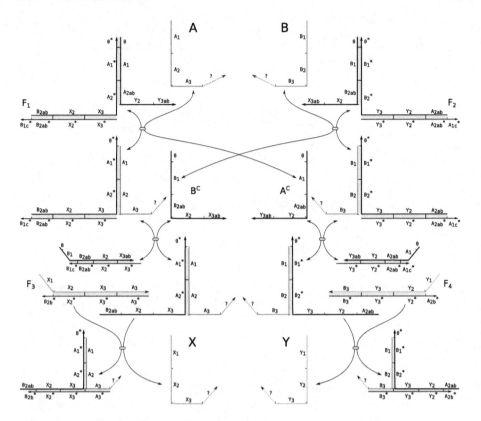

Fig. 9. An implementation of $A + B \rightarrow X + Y$ using the DLD motif and associative hybridization.

canonical reaction $A + B \rightarrow X + Y$ is illustrated in Fig. 8 and is a generalization of the cooperative AND gate implementation of Fig. 4. Input signals A and B cooperate to displace the long strand on fuel F_1, which in turn displaces two intermediate signals on F_2. One intermediate signal interacts with F_3 to produce output signal X, while the other interacts with F_4 to produce output signal Y. This process is driven forward since the final state contains four additional bound toeholds and maintains the same number of complexes as the initial state.

This reaction encoding is transactional. The process becomes irreversible only after the long strand on fuel F_1 is displaced by consuming both input signals. Thus, in the absence of one input the other cannot be irreversibly consumed. The remaining reactions produce the output species and render other species into waste.

6.2 DLD Motif + Associative Hybridization

Recall that cooperative strand displacement involves an initial effectively tri-molecular step. Such trimolecular reactions can be prohibitively slow for low concentration systems. We next demonstrate that arbitrary bimolecular chemi-cal reactions can in fact be implemented by utilizing the associative hybridization primitive. Although more complex, this scheme avoids the effectively trimolecu-lar step and thus scales more favorably to low concentrations.

An implementation of the canonical reaction $A+B \rightarrow X+Y$ using associative strand displacement is illustrated in Fig. 9. Consider how the product X, initially bound to fuel complex F_3, is produced. Two strand exchanges must occur on fuel complex F_1 to create a displacing complex with sufficient domains to displace the bound signal X. Firstly, input A exchanges with a "certificate" signal, A^C, which denotes that A has been consumed. The second exchange requires the certificate signal B^C, which denotes signal B has been consumed. After these strand exchanges, the domains X_2, X_3 and A_3 have been "glued" together (asso-ciated) and displacement of X can occur. Production of signal Y is symmetric, by design. This process is driven forward since the final state contains two addi-tional bound toeholds and maintains the same number of complexes as the initial state.

This reaction encoding is transactional. The process becomes irreversible only when the first output signal is produced (either X or Y). An output signal can only be produced after both input signals have been consumed. Once one output signal is produced, the other must eventually be produced since no series of backward reactions can occur in order to erroneously produce an input signal. (Suppose X has been produced. Then A is irreversibly bound and so is B^C. However, B^C is the only strand that could displace B).

7 Conclusion

The problem of leak has frustrated efforts to build complex DSD systems. In this work we begin a systematic effort to design DSD domain-level logic to reduce leak. In contrast to a number of previous approaches which relied on sequence-based leak reduction strategies, or subtle tweaks on existing designs (e.g. by introducing clamps), we rely on domain level redundancy. By utilizing more long domains in active signals, and more sequential strand displacement reactions to produce the output, we can increase the number of consecutive unfortunate events necessary for leak.

We focus on translator components, AND gates, and the implementation of CRNs. Our schemes rely on well-established types of strand displacement reac-tions, with the more complex components utilizing cooperative or associative displacement. It is, however, a natural open question whether it is possible to implement an NLD (or even DLD) AND gates and CRNs without using coop-erative or associative strand displacement.

We advance two types of arguments to affirm leak reduction. First, we obtain leak rates for specific constructions (DLD motif) using a domain level reaction enumerator. Second, we develop an analytical argument based on thermodynamics showing that increasing redundancy exponentially reduces the leak.

The principle remaining open questions concern the composition of the described components into circuits and networks. We have taken care to ensure that inputs and output signals have compatible form. However, we have not proven that the "leakless" properties of the individual components is preserved under composition. Further, properly ensuring the correctness of our CRN schemes when multiple reactions are implemented, even ignoring leak, requires more sophisticated arguments.

Acknowledgments. The authors are supported by a Banting Fellowship (CT), NSF CCF/HCC Grant No. 1213127, NSF CCF Grant No. 1317694, and NIGMS Systems Biology Center grant P50 GM081879 (DS). We thank Boya Wang and Robert Machinek for helpful discussions.

References

1. Cardelli, L.: Two-domain DNA strand displacement. Math. Struct. Comput. Sci. **23**(02), 247–271 (2013)
2. Chen, H.-L., Goel, A.: Error free self-assembly using error prone tiles. In: Ferretti, C., Mauri, G., Zandron, C. (eds.) DNA 2004. LNCS, vol. 3384, pp. 62–75. Springer, Heidelberg (2005)
3. Chen, X.: Expanding the rule set of DNA circuitry with associative toehold activation. J. Am. Chem. Soc. **134**(1), 263–271 (2011)
4. Chen, Y.-J., Dalchau, N., Srinivas, N., Phillips, A., Cardelli, L., Soloveichik, D., Seelig, G.: Programmable chemical controllers made from DNA. Nat. Nanotechnol. **8**(10), 755–762 (2013)
5. Dabby, N.L.: Synthetic molecular machines for active self-assembly: prototype algorithms, designs, and experimental study. Ph.D thesis, California Institute of Technology (2013)
6. Dong, Q.: A bisimulation approach to verification of molecular implementations of formal chemical reaction networks. Master's thesis, Stony Brook University (2012)
7. Genot, A.J., Bath, J., Turberfield, A.J.: Combinatorial displacement of DNA strands: application to matrix multiplication and weighted sums. Angew. Chem. Int. Ed. **52**(4), 1189–1192 (2013)
8. Grun, C., Sarma, K., Wolfe, B., Shin, S.W., Winfree, E.: The peppercorn enumerator. http://www.dna.caltech.edu/Peppercorn/
9. Grun, C., Sarma, K., Wolfe, B., Shin, S.W., Winfree, E.: A domain-level DNA strand displacement reaction enumerator allowing arbitrary non-pseudoknotted secondary structures. In: Verification of Engineered Molecular Devices and Programs (VEMDP) (2014). http://arxiv.org/abs/1505.03738
10. Jiang, Y.S., Bhadra, S., Li, B., Ellington, A.D.: Mismatches improve the performance of strand-displacement nucleic acid circuits. Angew. Chem. **126**(7), 1876–1879 (2014)
11. Krishnan, Y., Simmel, F.C.: Nucleic acid based molecular devices. Angew. Chem. Int. Ed. **50**(14), 3124–3156 (2011)
12. Lakin, M.R., Phillips, A., Stefanovic, D.: Modular verification of DNA strand displacement networks via serializability analysis. In: Soloveichik, D., Yurke, B. (eds.) DNA 2013. LNCS, vol. 8141, pp. 133–146. Springer, Heidelberg (2013)

13. Machinek, R.R.F.: Control and observation of DNA nanodevices. Ph.D thesis, University of Oxford (2014)
14. Phillips, A., Cardelli, L.: A programming language for composable DNA circuits. J. R. Soc. Interface **6**(Suppl 4), S419–S436 (2009)
15. Qian, L., Winfree, E.: Scaling up digital circuit computation with DNA strand displacement cascades. Science **332**(6034), 1196–1201 (2011)
16. Qian, L., Winfree, E., Bruck, J.: Neural network computation with DNA strand displacement cascades. Nature **475**(7356), 368–372 (2011)
17. Seelig, G., Soloveichik, D., Zhang, D.Y., Winfree, E.: Enzyme-free nucleic acid logic circuits. Science **314**(5805), 1585–1588 (2006)
18. Shin, S.W.: Compiling and verifying DNA-based chemical reaction network implementations. Master's thesis, California Institute of Technology (2011)
19. Soloveichik, D., Seelig, G., Winfree, E.: DNA as a universal substrate for chemical kinetics. Proc. Nat. Acad. Sci. **107**(12), 5393–5398 (2010)
20. Srinivas, N., Ouldridge, T.E., Šulc, P., Schaeffer, J.M., Yurke, B., Louis, A.A., Doye, J.P.K., Winfree, E.: On the biophysics and kinetics of toehold-mediated DNA strand displacement. Nucleic Acids Res. **41**(22), 10641–10658 (2013)
21. Teichmann, M., Kopperger, E., Simmel, F.C.: Robustness of localized DNA strand displacement cascades. ACS Nano **8**(8), 8487–8496 (2014)
22. Winfree, E., Bekbolatov, R.: Proofreading tile sets: error correction for algorithmic self-assembly. In: Chen, J., Reif, J. (eds.) DNA Computing. LNCS, vol. 2943, pp. 126–144. Springer, Heidelberg (2004)
23. Yin, P., Choi, H.M.T., Calvert, C.R., Pierce, N.A.: Programming biomolecular self-assembly pathways. Nature **451**(7176), 318–322 (2008)
24. Zhang, D.Y.: Cooperative hybridization of oligonucleotides. J. Am. Chem. Soc. **133**(4), 1077–1086 (2010)
25. Zhang, D.Y., Seelig, G.: Dynamic DNA nanotechnology using strand-displacement reactions. Nat. Chem. **3**(2), 103–113 (2011)
26. Zhang, D.Y., Turberfield, A.J., Yurke, B., Winfree, E.: Engineering entropy-driven reactions and networks catalyzed by DNA. Science **318**(5853), 1121–1125 (2007)
27. Zhang, D.Y., Winfree, E.: Control of DNA strand displacement kinetics using toehold exchange. J. Am. Chem. Soc. **131**(47), 17303–17314 (2009)

Supervised Learning in an Adaptive DNA Strand Displacement Circuit

Matthew R. Lakin[1,2]([✉]) and Darko Stefanovic[1,2]

[1] Department of Computer Science, University of New Mexico,
Albuquerque, NM 87131, USA
[2] Center for Biomedical Engineering, University of New Mexico,
Albuquerque, NM 87131, USA
{mlakin,darko}@cs.unm.edu

Abstract. The development of DNA circuits capable of adaptive behavior is a key goal in DNA computing, as such systems would have potential applications in long-term monitoring and control of biological and chemical systems. In this paper, we present a framework for adaptive DNA circuits using buffered strand displacement gates, and demonstrate that this framework can implement supervised learning of linear functions. This work highlights the potential of buffered strand displacement as a powerful architecture for implementing adaptive molecular systems.

1 Introduction

Implementing adaptive behaviors, such as supervised learning, is a key challenge for the fields of molecular computing and synthetic biology. Addressing this challenge would enable the development of molecular computing solutions to important practical applications, such as the detection of emerging pathogens [1] whose signatures mutate over time. Furthermore, the development of molecular computing systems that are capable of operating over an extended period of time would advance the state of the art of molecular circuit design, as most current systems are single-use devices.

Previous experimental work has shown that neural networks may be implemented using DNA strand displacement circuits [2] comprising "seesaw" gates [3,4]. However, in that work the neural network was trained *in silico* and each instance of the experimental system could only be used one time. Our prior theoretical work has demonstrated that a biochemical system assuming hypothetical, DNAzyme-like reactions can learn a class of linear functions [5], and other work has shown that high-level artificial chemistries can learn to implement Boolean functions [6] and perceptron-like classification tasks [7]. Here we present a design for an adaptive, reusable DNA learning circuit based on the framework of four-domain DNA strand displacement reactions, which has been shown to be capable of implementing arbitrary chemical reaction networks in a concrete biochemical system [8]. Thus, this paper offers a route to an experimental realization of adaptive DNA computing systems.

The remainder of this paper is organized as follows. In Sect. 2, we introduce buffered DNA strand displacement systems and illustrate their use for

© Springer International Publishing Switzerland 2015
A. Phillips and P. Yin (Eds.): DNA 2015, LNCS 9211, pp. 154–167, 2015.
DOI: 10.1007/978-3-319-21999-8_10

implementing adaptive systems. In Sect. 3, we present an adaptive buffered amplifier, which is a key component of the learning circuit that we present in Sect. 4. We present results from computational simulations of this learning circuit in Sect. 5, and conclude with a discussion in Sect. 6.

2 Buffered DNA Strand Displacement for Adaptive Systems

The idea of buffered DNA strand displacement gates was introduced (as "curried" gates) by Cardelli [9]. This idea was further developed in the context of implementing DNA oscillators with robust long-term kinetics [10]. In this section we recap the principle of buffered strand displacement gates and show how they can be used to implement adaptive molecular systems.

The basic principle of buffered strand displacement is illustrated in Fig. 1. In this paper we base our gate designs on the "four-domain" encoding of abstract chemical reaction networks into DNA strand displacement, developed by Soloveichik [8]. In the four-domain encoding, each abstract species X is represented by three domains $X\hat{a}$, Xb and $X\hat{c}$. We write "?" to denote a "history domain" whose identity is irrelevant for the operation of the gate (however, the history domains for the product species must be freshly generated for each gate, to avoid crosstalk). The key difference is that, rather than initializing the system with populations of active gates capable of accepting input strands directly, a buffered system is initialized with a *buffer* of inactive gates that cannot initially accept input strands. A population of "unbuffering" strands must first be introduced, which (irreversibly) activate a subset of the buffered gates, so that they can accept input strands as normal. We design the gates so that an additional copy of each gate's unbuffering strand is released along with the gate's outputs, to maintain a (roughly) constant population of active gates. Furthermore, since the gates in the buffer are inactive, the population of inactive buffered gates can be replenished, either at intervals or continuously, without significantly affecting the reaction kinetics. If waste products were removed, to prevent them from accumulating, this could allow buffered strand displacement systems to run indefinitely.

We will write a buffered strand displacement gate that implements the reaction $\overline{R} \rightarrow \overline{P}$ with buffer species B using the notation $B \vdash \overline{R} \rightarrow \overline{P}$, and we will use the $\varnothing$ symbol to denote an empty (multi)set of reactants or products.

We propose to use buffered strand displacement gates to implement adaptive systems. To this end, we note that altering the concentration of unbuffering strands that are available for a given buffered gate provides a means of controlling the rate of the gate's reaction: if more copies of a given reaction gate are activated from the buffer, it will emulate a faster reaction. (This is a simpler approach than engineering toehold binding energies or developing remote toehold systems [11].) Furthermore, in addition to regenerating its own unbuffering strand, a given buffered gate may also release unbuffering strands for other gate populations as part of its output. This crucial fact enables one buffered gate to adjust the number of active gates of a second kind, thereby controlling the rate of the second buffered gate's reaction.

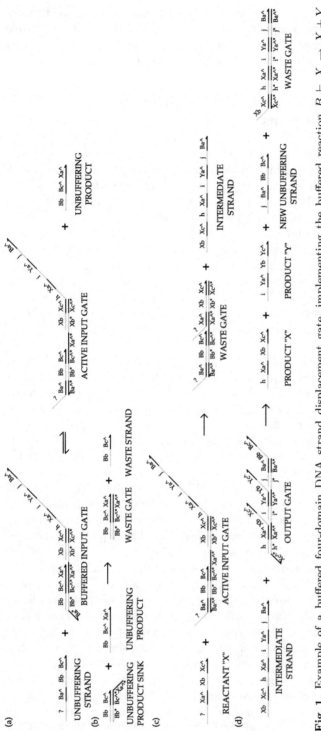

Fig. 1. Example of a buffered four-domain DNA strand displacement gate, implementing the buffered reaction $B \vdash X \rightarrow X + Y$. **a.** The first reaction is the unbuffering reaction, in which an unbuffering strand binds to a buffered gate and reversibly produces an active input gate. The binding toehold for the X reactant strand is exposed in the active input gate. **b.** To make the unbuffering reaction effectively irreversible, an additional sink gate for the strand released by the unbuffering reaction is provided (this gate is not required in standard four-domain chemical reaction network encodings). **c.** A reactant strand X can bind to an active input gate via the $X\hat{a}$ toehold, irreversibly displacing an intermediate strand. **d.** The released intermediate strand binds to the output gate and releases all of the products in a single strand displacement reaction. Here, in addition to the product strands X and Y, an additional output, B, is generated, which is a new copy of the unbuffering strand for this gate. Thus, a new gate will be activated from the buffer to replace the gate that was consumed to execute this reaction.

3   An Adaptive, Buffered Amplifier

In this section, we will illustrate the use of buffered strand displacement gates to implement an adaptive amplifier whose gain can be dynamically adjusted. This contrasts with previous work on DNA strand displacement-based amplifiers [3,4,12–14], which relied on hard-coding the gain of the system in the initial species concentrations, and gave no consideration to reusability or autonomously adjusting the gain of the amplifier. In addition to providing an example of a buffered strand displacement system whose result can be controlled by adjusting the provision of unbuffering strands, this circuit design motif will be a key component of our learning circuit design.

Our design for a reusable, adaptive, buffered strand displacement-based amplifier consists of two buffered strand displacement gates:

$$B \vdash x \rightarrow x + y \qquad\qquad B' \vdash x \rightarrow \varnothing.$$

The first gate is a "catalyst" gate that uses the input strand x to catalyze release of the output strand y, and the second gate is a "degradation" gate that removes the input strand x from the system by consuming it without releasing

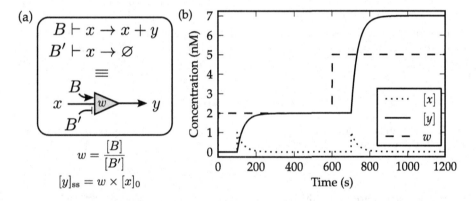

Fig. 2. Buffered amplifier design and ODE simulation data. **a.** The buffered amplifier consists of two buffered reactions: a catalytic reaction that generates the output and a degradation reaction that removes the input from the system. The ratio of the concentrations of active gates from the two buffers corresponds to the gain of the amplifier. **b.** Data from an ODE simulation of the buffered amplifier with multiple sequential input additions and dynamic control of amplifier gain. The initial concentration of B' was 100 nM and the initial concentration of B was 200 nM, which sets the gain, w, to be 2.0. Addition of the x input species (1 nM) at $t = 100$ s causes the amplifier to generate the output species y at a concentration of 2 nM, as expected. At $t = 600$ s, an additional 300 nM of the unbuffering strand B was added, increasing the amplifier's gain (w) by 3.0. Then, subsequent addition of x (1 nM) at $t = 700$ s produced a further 5 nM of the output species y, showing that the gain had indeed been increased to 5.0, as expected. (Note that the plotted value of w is not a concentration but rather a ratio of concentrations, but the value follows the left axis.)

any output species (except for a new activating strand B'). This design draws on ideas introduced by Zhang and Seelig [14]: the ratio between the effective rates of the first and second gates controls the number of output molecules y that each input molecule x can produce before it is irreversibly consumed, and therefore this ratio controls the gain of the amplifier. Thus, in the amplifier circuit, the concentration ratio $\frac{[B]}{[B']}$ controls the gain of the amplifier. More specifically, if we write $[z]_0$ for the initial concentration of species z and $[z]_{ss}$ for the steady state concentration of z (assuming that a steady state exists), then we would expect that $[y]_{ss} = \frac{[B]_0}{[B']_0} \times [x]_0$. A graphical shorthand for this circuit design element is presented in Fig. 2(a). Our approach to implementing the amplifier circuit motif is also related to the "ideal gain blocks" developed by Oishi and Klavins [15], the main difference being that our approach produces a quiescent final state, whereas their approach produces a steady state in which production and degradation of the output species balance. Thus, our approach prevents the supply of input species being constantly drained as in [15]. However, our approach does require some care to be taken when composing the output from an amplifier circuit motif with downstream circuit elements, as we outline below.

We implemented this circuit motif in the buffered strand displacement gate framework introduced in Sect. 2, and simulated it using the beta version of the DSD compiler that includes support for mixing events [16]. These simulation results are presented in Fig. 2(b), and show that the amplifier produces correct results for several gain settings, and that the gain of the amplifier can be dynamically adjusted by directly adding more unbuffering strands between amplification reactions.

4   A Strand Displacement Learning Circuit

In this section, we will present a buffered strand displacement system that can learn linear functions f of the form

$$f(x_1, x_2) = w_1 \times x_1 + w_2 \times x_2,$$

where w_1 and w_2 are real-valued coefficients and x_1 and x_2 are real-valued inputs. In the current paper, we restrict ourselves to the two-input case for clarity, although the circuit design motifs that we will present could be replicated to handle more inputs. In particular, a bias term could be included by incorporating an additional input signal x_0 that is always supplied with the input value $x_0 = -1$ in each training round. This is standard practice in studies of artificial neural networks. We will present a strand displacement system that learns functions of this form using a stochastic gradient descent algorithm [17]. Gradient descent is a general solution for many optimization tasks, in which the current weight approximations are adjusted to minimize the squared error over the entire training set in each training round. Stochastic gradient descent is a simplification of gradient descent that only considers a single training instance in each training round, making it more amenable to implementation in a molecular computing system.

A molecular computing system that solves this problem must accept the input values x_1 and x_2, compute the predicted output $y = \widehat{w_1} \times x_1 + \widehat{w_2} \times x_2$ based on the current stored weight approximations $\widehat{w_1}$ and $\widehat{w_2}$, compare y with the supplied expected value $d = w_1 \times x_1 + w_2 \times x_2$, and update the stored weight approximations according to the gradient descent weight update rule:

$$\widehat{w_i} := \widehat{w_i} + \alpha \times (d - y) \times x_i, \tag{1}$$

where α is a (small) positive coefficient called the "learning rate". We now present a strand displacement system that implements this learning algorithm using buffered DNA strand displacement gates. Our design can be divided into two subcircuits, as follows.

Predictor Subcircuit. The predictor subcircuit is based on the adaptive amplifier presented in Sect. 3. The input signals, and all other numeric signals, are represented in a dual rail format with a differential encoding, that is, the input signal x_i actually consists of two signals x_i^+ and x_i^-, and the value of x_i is interpreted as the concentration difference $[x_i^+] - [x_i^-]$. The predictor subcircuit design is presented in Fig. 3: the initial circuit motif is replicated for each input x_i. Here, and henceforth, we will omit the identities of buffer species from figures if their identity is not important when describing the operation of the circuit. The species highlighted in grey (x_i^+, x_i^-, d^+, and d^-) are provided by the user to initiate each training round: their concentrations represent the input value x_i and the expected result d that the user must derive using the target weight values. (If $d = 0$ then we must add equal concentrations of d^+ and d^-. Any non-zero concentration is acceptable, as these species must be present to drive the execution of the predictor subcircuit).

Each positive input signal x_i^+ is "copied" by a buffered fork gate that generates four signals with the same overall concentration as the original: two of these, x_{i1}^+ and x_{i2}^+, are for use by the predictor subcircuit and the remaining two, k_{i1}^+ and k_{i2}^+, are for use by the feedback subcircuit (as detailed below). Each negative input signal x_i^- is copied similarly.

The key parts of the predictor subcircuit are the buffered strand displacement amplifier motifs. The initial gains of the predictor subcircuit amplifiers encode the initial approximation of each weight value stored in the system. There is one pair of amplifiers per positive input signal and one pair per negative input signal. In each of these pairs, the gain of one amplifier represents the positive component $\widehat{w_i}^+$ of the corresponding weight approximation $\widehat{w_i}$, and the gain of the other amplifier represents the negative component $\widehat{w_i}^-$. Thus, the positive component of each input is multiplied by both the positive and negative components of the corresponding weight, and similarly for the negative component. The outputs of the predictor subcircuit amplifiers are two species y^+ and y^-, which represent positive and negative components of the current prediction, based on the current input values and stored weight approximations. The amplifier gates are constructed such that the sign of the output species is correct with respect to the signs of the input component and the weight component in each case.

To complete the execution of the predictor subcircuit, the y^+ and y^- species, whose concentrations represent the current prediction, interact with the d^+ and d^- species, whose concentrations represent the expected value of the target function. These species interact via the four buffered reactions shown in the box on the right-hand side of Fig. 3, which collectively have the effect of subtracting the value of y from the value of d. We implement this operation via four two-input two-output reactions, in which the $d^\pm$ species catalyze conversion of the $y^\pm$ species to $d^\pm$, with signs chosen such that the resulting concentrations of $d^\pm$ represent the result of a subtraction. We choose to implement subtraction in this way, rather than using annihilator gates that degrade the positive and negative variants of the species, to avoid sequestration of the remaining species by the annihilator gates, which has been problematic in other work [16]. For this reason, it is crucial that the $y^\pm$ species are the first input strands consumed by these reaction gates.

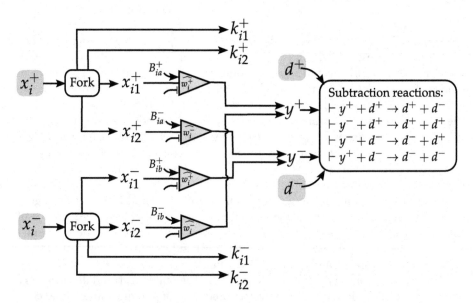

Fig. 3. Design schematic for the predictor subcircuit. The concentrations of the input species $x_i{}^+$ and $x_i{}^-$ (for each input signal x_i) are copied by buffered fork gates to produce species that serve as inputs to the buffered amplifier motifs that implement linear function prediction, and additional species ($k_{i1}{}^\pm, k_{i2}{}^\pm$) that will be used in the feedback subcircuit. The gains of these amplifiers store the current weight approximations. The amplifiers produce species y^+ and y^-, such that $[y^+] - [y^-]$ equals the predicted output value. These species then interact with the d^+ and d^- species, which encode the expected output value, via four buffered reactions that implement subtraction. When the predictor subcircuit reaches steady state, the concentration difference $[d^+] - [d^-]$ should equal $d - y$, i.e., the error in the prediction when compared with the expected output value.

Thus, when the predictor subcircuit reactions reach steady state, the concentration difference $[d^+] - [d^-]$ represents the value of $d - y$. If this value is positive, i.e., if $[d^+] > [d^-]$, then the predicted output value was too small. Similarly, if this value is negative, i.e., if $[d^+] < [d^-]$, then the predicted output was too large. The goal of the learning process is to adjust the stored weight approximations

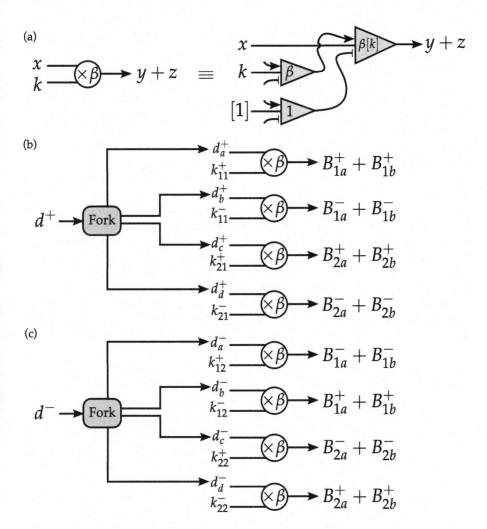

Fig. 4. Feedback subcircuit. **a.** Graphical shorthand for a multi-amplifier motif that enables one input concentration ($[x]$) to be multiplied by another concentration ($[k]$) and a scalar scaling factor (β). **b.** and **c.** Design schematic for the feedback subcircuit. Here, the scaling factor $\beta = \alpha \times \delta$, where α is the learning rate constant and δ is the denominators of the weight ratios in the predictor subcircuit. The feedback circuitry from **b.** is activated when d^+ is left over from the predictor subcircuit, and the feedback circuitry from **c.** is activated when d^- is left over from the predictor subcircuit.

$\widehat{w_i}$ to match the target weight approximations, so that $[d^+] = [d^-]$ when the predictor subcircuit reactions reach steady state.

Feedback Subcircuit. Once the predictor subcircuit has computed the discrepancy between the predicted function output and the expected output, the feedback subcircuit must use the value of this discrepancy to update the weight approximations stored in the predictor subcircuit, according to the gradient descent learning rule (1). A design schematic for our feedback subcircuit is presented in Fig. 4.

The first point to note from (1) is that the feedback subcircuit must take the concentration of $d^\pm$ that denotes the discrepancy from the predictor subcircuit and, for each input, multiply the discrepancy value by the corresponding input value, both of which are concentrations, and by the learning rate constant α. However, a single buffered amplifier can only multiply an input concentration by a gain factor encoded as a ratio of concentrations. To enable two input concentrations to be multiplied together, we have developed a two-concentration multiplier circuit motif, shown in Fig. 4(a). In this motif, we assume that no unbuffering strands are initially present for the uppermost amplifier, which will accept the input signal x and produce the output species. The input signal k activates an amplifier that produces the unbuffering strand for the catalyst gate from the output-producing amplifier, with gain β. An additional input signal, whose value is constant 1, activates an amplifier that produces the unbuffering strand for the degradation gate from the output-producing amplifier, with gain 1. Thus, these secondary amplifiers preset the gain of the output-producing amplifier to be $\beta \times [k]$, and upon addition of the input signal x the resulting concentration of both output species (y and z) will be $\beta \times [x] \times [k]$, as desired. (Here, we include two outputs because this variation is called for in our two-input learning circuit design.)

The feedback subcircuit uses the two-concentration multiplier circuit motif extensively, to implement the weight update rule (1). Figure 4(b)(c) presents the feedback subcircuit design for the two-input case. Execution of the feedback subcircuit is initialized by buffered fork gates that copy the $d^\pm$ species into eight species $d_a{}^\pm$, $d_b{}^\pm$, $d_c{}^\pm$, and $d_d{}^\pm$. We assume that there are initially no unbuffering strands for these fork gates and, once the predictor subcircuit has run to completion, the addition of these unbuffering strands triggers execution of the feedback subcircuit.

For each combination of signs for the leftover $d^\pm$ species and the copied input species $k_{ij}{}^\pm$ from the predictor subcircuit, the copied $d^\pm$ species and the copied $k_{ij}{}^\pm$ species serve as inputs to an instance of the two-concentration multiplier circuit motif. As described above, this circuit motif multiplies these values together (and by a scaling factor $\beta = \alpha \times \delta$, where α is the learning rate constant and δ is the denominators of the weight ratios in the predictor subcircuit) and generates additional unbuffering strands for certain amplifier gates from the predictor subcircuit. From the two-input predictor subcircuit design from Fig. 3, we see that additional unbuffering strands B_{ia}^+ and B_{ib}^+ must be generated to increase weight approximation $\widehat{w_i}$, and that additional unbuffering strands B_{ia}^- and B_{ib}^-

must be generated to decrease weight approximation $\widehat{w_i}$. These pairs of species must be generated together, so that the pairs of amplifiers from the predictor subcircuit that are initialized with the same weight approximation are updated together.

5   Results

We encoded the learning circuit design from Sect. 4 in the DSD programming language [18] and used the associated DSD compiler, with the *"Infinite"* reaction semantics [10], to generate MATLAB code that implements an ODE model of the system. The initial state of the two-input system consists of 278 species, of which the majority (222) are gate complexes. It is worth noting, however, that many of these species are variants with different combinations of history domains, so the design complexity of the circuit is not as high as suggested by the raw species counts. Furthermore, the number of species should scale linearly with the number of input signals, so the circuit design presented here could be replicated to learn similar functions of more than two inputs.

We simulated the ODE model of the learning circuit using a custom MAT-LAB simulation routine that invokes a stiff ODE solver and that also allows mixing events to be executed at certain time points during the simulation. These mixing events simulate the addition of inputs to the system, or the removal of species, by the experimenter at certain time points. In principle, stochastic simulations could also be used to investigate the behavior of the system in the limit of low species populations, though the requirement for high populations of buffered gates could lead to poor performance in a stochastic simulation of the full system. We found that the size of the buffer, that is, the quantity of unbuffered gates waiting to be activated, did not affect the accuracy of the computation performed by the learning circuit. It did, however, reduce the number of training rounds that could be conducted before the buffer was depleted, at which point no further training could be carried out. This point is discussed in Sect. 6 below.

The initial state of the system consists of the various buffered gates and their unbuffering strands (with the exceptions of the unbuffering strands for the fork gates and output-generating amplifiers in the feedback subcircuit, as described above). The initial weight approximations $\widehat{w_i}$ are encoded as the gain settings of the amplifiers in the predictor subcircuit. After the gates have unbuffered (after 500 s), the first training inputs are added, which consist of the $x_i^{\pm}$ species and the $d^{\pm}$ species, whose concentrations encode the input values and the expected function output, respectively. At the same time, we also add the constant-valued signals that serve as an input to the feedback subcircuit, so that the output-generating amplifiers from the feedback subcircuit will be primed with the correct gain values before it starts executing. After a further 2000 s, when the predictor subcircuit has completed its execution, we add unbuffering strands for the fork gates from the feedback subcircuit, which triggers execution of the feedback subcircuit. After a further 3500 s, when the feedback subcircuit has finished updating the weight approximations stored in the predictor subcircuit, we set the concentrations of any remaining unbuffered (active) fork gates and output-generating gates in the feedback subcircuit to zero, to reset the state of the feedback subcircuit. We also add the second set of training inputs at this time, and iterate until the specified sequence of training instances have all been presented.

We ran a total of 1,000 simulations, with initial and target weight values and input values selected from a uniform random distribution over the interval $[-10, 10]$ and with a fixed learning rate value $\alpha = 0.01$. Figure 5 shows simulation results illustrating how the weight values evolve over the course of two 15-round example training schedules. In the two-input case, we can plot the weight space as a 2D plot and observe the trajectory of how the weights evolve from their initial values (labeled "START") towards the true values (labeled "TARGET") over the training period. In each example, the trajectory for the simulation of the DNA strand displacement learning circuit is overlaid with a trajectory derived from a reference implementation of the stochastic gradient descent learning rule (1) in Python, using the same initial state and the same training schedule. The weight trajectories shown in Fig. 5 are representative of the agreement between the DNA system and the reference implementation observed in all cases, and of the fact that the weight trajectories correctly home in on the target weight values. These results give us confidence in the correct operation of our circuit design as an implementation of the learning rule (1).

Furthermore, to investigate the aggregate learning performance from our simulations, we computed the root mean square error (RMSE) of the stored weight approximations compared with the target weight values at each step of each of the 1,000 training simulations. The average RMSE values over the 1,000 training simulations, and the standard deviations, are plotted in Fig. 6. Again, the results from the DNA and Python versions of the learning system are overlaid precisely, indicating that the DNA circuit follows the expected behavior.

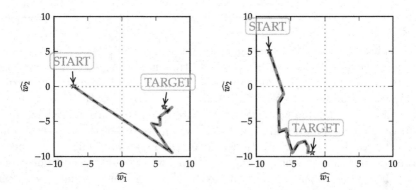

Fig. 5. Two example weight traces from DNA learning circuit simulations (grey solid lines), overlaid with corresponding traces from our Python reference implementation (black broken lines). The lines coincide closely in all simulations, not just the examples shown here, indicating that the DNA circuit works correctly.

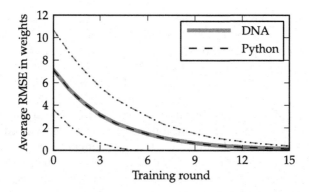

Fig. 6. Learning curves from DNA learning circuit simulations and our Python reference implementation. The average RMSE in the weight approximations was computed after each training round for 1,000 training schedules, each 15 rounds in length. Non-solid lines are one standard deviation above and below the mean. Again, the lines coincide very well, indicating that our DNA circuit design works correctly.

The RMSE in the weight approximations is almost zero after just 15 training rounds, suggesting that an experimental implementation of this system could be trained to a reasonable degree of accuracy in a limited timeframe.

6   Discussion

We have presented a design for a DNA learning circuit based on buffered DNA strand displacement reactions, and demonstrated via simulation results that the design works as intended and can learn target weight values in a reasonable timeframe. This feedback subcircuit design surpasses our previous work in this area [5], by allowing negative weight values and input values. It also alleviates the problems we saw in [5] with poor performance when trying to learn weight values near zero, by implementing a weight update rule that is symmetric in the positive and negative directions. The weight update rule used in this work (1) is a well-studied gradient descent learning scheme, whose performance has been studied extensively [17]. It is well known that the learning rate parameter (α) can have a significant effect on the rate of convergence of the learning process. Indeed, many different procedures for reducing the learning rate over time to achieve rapid convergence have been studied, and our system should respond to such adjustments in the same way as the reference algorithm that we implemented in Python.

Our circuit was specifically designed to avoid the issue of input sequestration, which can be problematic in DNA strand displacement systems. Soloveichik's compilation scheme for abstract chemical reactions [8] dealt with this issue by adding additional reaction gates to compensate by sequestering all other species similarly, which slows down the system. Other work [16] has dealt with input sequestration by artificially increasing the values of certain input signals. This consideration led to the design of the "subtractor" circuit motif shown in Fig. 3 which, while elegant, has the property that the absolute values of the positive and negative components of the weight approximations and other signals processed by the system grow monotonically over the course of a multi-round training session, draining other species out of the system. This problem could be ameliorated by the inclusion of an annihilator gate in the feedback subcircuit that takes d^+ and d^- as inputs but produces no output, which could reduce the absolute sizes of the signals. This annihilator gate would only need to compete with the rest of the feedback subcircuit to drain some of the excess $d^{\pm}$ signals from the system, and we will explore this alternative in future work.

Our use of buffered gates to implement adaptive strand displacement circuits offers a route to implement systems whose operation can be extended by the replenishment of buffered gate species when they are depleted, without adversely affecting the kinetics of the system. This is an important consideration for the implementation of molecular learning circuits. Furthermore, in this paper we have implemented the stochastic gradient descent weight update rule in conjunction with a linear transfer function, which allows the circuit to learn linear classification functions. However, the feedback subcircuit design presented here could be used to learn other functional forms, including non-linear functions, by simply replacing the predictor subcircuit to compute the desired function of the provided training inputs. A major challenge is to implement other transfer functions, in particular, non-linear transfer functions such as the Heaviside function, which is used in classical expositions of perceptron learning [19]. Specifically, the challenge here is to implement a reusable circuit that can amplify its output to a fixed level. Finally, to build molecular systems that can learn arbitrary functions it would be necessary to connect a number of such units into networks to be trained by backpropagation, which could be achieved by cascading several of the circuit motifs described here.

Acknowledgments. This material is based upon work supported by the National Science Foundation under grants 1318833 and 1422840. M.R.L. gratefully acknowledges support from the New Mexico Cancer Nanoscience and Microsystems Training Center.

References

1. Morens, D.M., Fauci, A.S.: Emerging infectious diseases: threats to human health and global stability. PLOS Pathog. **9**(7), e1003467 (2013)
2. Zhang, D.Y., Seelig, G.: Dynamic DNA nanotechnology using strand-displacement reactions. Nat. Chem. **3**(2), 103–113 (2011)
3. Qian, L., Winfree, E.: A simple DNA gate motif for synthesizing large-scale circuits. J. R. Soc. Interface **8**(62), 1281–1297 (2011)
4. Qian, L., Winfree, E., Bruck, J.: Neural network computation with DNA strand displacement cascades. Nature **475**, 368–372 (2011)
5. Lakin, M.R., Minnich, A., Lane, T., Stefanovic, D.: Design of a biochemical circuit motif for learning linear functions. J. R. Soc. Interface **11**(101), 20140902 (2014)

6. Banda, P., Teuscher, C., Lakin, M.R.: Online learning in a chemical perceptron. Artif. Life **19**(2), 195–219 (2013)
7. Banda, P., Teuscher, C., Stefanovic, D.: Training an asymmetric signal perceptron through reinforcement in an artificial chemistry. J. R. Soc. Interface **11**, 20131100 (2014)
8. Soloveichik, D., Seelig, G., Winfree, E.: DNA as a universal substrate for chemical kinetics. Proc. Natl. Acad. Sci. USA **107**(12), 5393–5398 (2010)
9. Cardelli, L.: Strand algebras for DNA computing. Nat. Comput. **10**(1), 407–428 (2010)
10. Lakin, M.R., Youssef, S., Cardelli, L., Phillips, A.: Abstractions for DNA circuit design. J. R. Soc. Interface **9**(68), 470–486 (2012)
11. Genot, A.J., Zhang, D.Y., Bath, J., Turberfield, A.J.: Remote toehold: a mechanism for flexible control of DNA hybridization kinetics. J. Am. Chem. Soc. **133**, 2177–2182 (2011)
12. Zhang, D.Y., Turberfield, A.J., Yurke, B., Winfree, E.: Engineering entropy-driven reactions and networks catalyzed by DNA. Science **318**, 1121–1125 (2007)
13. Qian, L., Winfree, E.: Scaling up digital circuit computation with DNA strand displacement cascades. Science **332**, 1196–1201 (2011)
14. Zhang, D.Y., Seelig, G.: DNA-based fixed gain amplifiers and linear classifier circuits. In: Sakakibara, Y., Mi, Y. (eds.) DNA 16 2010. LNCS, vol. 6518, pp. 176–186. Springer, Heidelberg (2011)
15. Oishi, K., Klavins, E.: Biomolecular implementation of linear I/O systems. IET Syst. Biol. **5**(4), 252–260 (2011)
16. Yordanov, B., Kim, J., Petersen, R.L., Shudy, A., Kulkarni, V.V., Phillips, A.: Computational design of nucleic acid feedback control circuits. ACS Synth. Biol. **3**(8), 600–616 (2014)
17. Duda, R.O., Hart, P.E., Stork, D.G.: Pattern Classification, 2nd edn. Wiley, New York (2001)
18. Lakin, M.R., Youssef, S., Polo, F., Emmott, S., Phillips, A.: Visual DSD: a design and analysis tool for DNA strand displacement systems. Bioinformatics **27**(22), 3211–3213 (2011)
19. Minsky, M., Papert, S.: Perceptrons: an introduction to computational geometry, 2nd edn. MIT Press, Cambridge (1972)

Automated Design and Verification
of Localized DNA Computation Circuits

Michael A. Boemo[1]([✉]), Andrew J. Turberfield[1], and Luca Cardelli[2,3]

[1] Department of Physics, Clarendon Laboratory, University of Oxford,
Parks Road, Oxford OX1 3PU, UK
{michael.boemo,a.turberfield1}@physics.ox.ac.uk
[2] Department of Computer Science, Wolfson Building, University of Oxford,
Parks Road, Oxford OX1 3QD, UK
[3] Microsoft Research Cambridge, Station Road, Cambridge CB1 2FB, UK
luca@microsoft.com

Abstract. Simple computations can be performed using the interactions between single-stranded molecules of DNA. These interactions are typically toehold-mediated strand displacement reactions in a well-mixed solution. We demonstrate that a DNA circuit with tethered reactants is a distributed system and show how it can be described as a stochastic Petri net. The system can be verified by mapping the Petri net onto a continuous time Markov chain, which can also be used to find an optimal design for the circuit. This theoretical machinery can be applied to create software that automatically designs a DNA circuit, linking an abstract propositional formula to a physical DNA computation system that is capable of evaluating it.

Computation with DNA has been the subject of much interest from the points of view of both pure computer science and nanomedicine. A 2009 paper by Andrew Phillips and Luca Cardelli showed how DNA strand displacement can be thought of as a formal computing language [8]. Further work by Matthew Lakin and colleagues produced Microsoft Visual DSD, a computational tool for the design and analysis of such reactions [6]. In the field of nanomedicine, Benenson et al. created a biomolecular DNA computing system that can produce an mRNA inhibitor to control gene expression [2].

These papers all consider DNA strands as freely floating reactants in a well-mixed solution. There are no topological or geometric constraints that prevent two species from interacting. Such constraints can be introduced by tethering DNA reactants to rigid structures. Yin and colleagues designed a DNA Turing machine that operates by DNA walkers moving on a rigid lattice [11]. Another method utilizes the tethering of walkers to a DNA origami tile [9]. In a 2005 paper, Jonathan Bath and colleagues introduced a DNA walker powered by a nicking enzyme that is capable of traversing a track of single-stranded DNA anchorages (Fig. 1) [1]. Shelley Wickham and colleagues built on this design in a 2011 paper that demonstrated how the walker could be programmed to navigate a series of tracks on an origami tile [10]. The result was a DNA walker that

© Springer International Publishing Switzerland 2015
A. Phillips and P. Yin (Eds.): DNA 2015, LNCS 9211, pp. 168–180, 2015.
DOI: 10.1007/978-3-319-21999-8_11

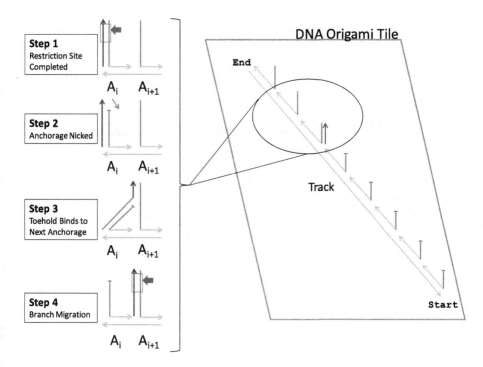

Fig. 1. The "burnt-bridges" DNA walker from [1]. Single-stranded oligonucleotides are shown as arrows, with the arrowhead indicating the 3' end. *Step 1* A DNA walker (shown in blue) is bound to a complementary single-stranded anchorage (A_i), completing the restriction site (red box) for a nicking enzyme. *Step 2* The nicking enzyme cuts the 5' end of anchorage A_i, exposing a 6-nucleotide toehold on the 3' end of the DNA walker. *Step 3* The toehold on the walker binds to the next anchorage in the track, anchorage A_{i+1}. *Step 4* The DNA walker moves from anchorage A_i to A_{i+1} via a toehold-mediated branch migration reaction. By continuously repeating Steps 1–4, a DNA walker can navigate down a track of single-stranded anchorages.

could perform a computation, namely a binary decision tree. This is a "local" computation that is performed by reactants that are tethered to the origami tile.

Localized DNA computation has also been the subject of theoretical and computational study. A recent paper by Dannenberg et al. analyzed the computational potential of localized DNA circuits [3]. Lakin and colleagues incorporated tethered DNA reactants into Microsoft Visual DSD, allowing for topological and geometric constraints in DNA circuits [7]. This software brings the functionality of Visual DSD to localized DNA computation circuits: It can perform probabilistic model checking, detect leaks, and provide information about reaction rates.

We demonstrate how localized DNA circuits can be analyzed as distributed systems. As such, they can be automatically designed and verified by software. The input to the localized DNA circuit can be posed using the language and grammar of the propositional calculus (Sect. 1). This input is then abstracted into a directed graph that captures the topology of the circuit (Sect. 2). Because the localized DNA circuit is a distributed system, it can be modeled as a stochastic Petri net. Analysis of this Petri net determines the geometry of the circuit (Sect. 3).

1   The Propositional Calculus of DNA Localized Computation Circuits

The language of a propositional calculus is composed of two parts: The first is a set of propositional variables, or atomic statements that each hold exactly one truth value (1 or 0); the second is the set of logical connectives, or operators that act on the propositional variables. A propositional formula is a string of propositional variables and logical connectives that is said to be well-formed if it follows the rules of the grammar.

Localized DNA computation systems can be designed to evaluate propositional formulae, and their action can be written in the formal language and grammar of the propositional calculus. The set of all logical connectives Ω can be partitioned into disjoint subsets according to their arity, or the number of arguments each connective takes:

$$\Omega = \Omega_0 \cup \Omega_1 \cup \Omega_2 \cup \cdots \cup \Omega_n.$$

For localized DNA computation systems, attention is restricted to nullary, unary, and binary logical connectives where,

$$\Omega_0 = \{\mathbf{0}, \mathbf{1}\},$$
$$\Omega_1 = \{\neg\},$$
$$\Omega_2 = \{\vee, \wedge, \rightarrow, \leftrightarrow\}.$$

The rules of the propositional calculus can be used to search for a simpler form of a propositional formula in order to make the corresponding DNA computation system more efficient. For example, the propositional formula

$$(x \wedge y) \vee (x \wedge z) \tag{1}$$

has three logical connectives, and hence will require three logic gates. In this case, it is possible to find an equivalent form that requires only two:

$$(x \wedge y) \vee (x \wedge z) \equiv x \wedge (y \vee z). \tag{2}$$

There are libraries available that can implement heuristics for such a search, e.g. SymPy [4].

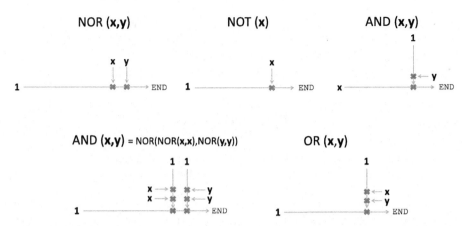

Fig. 2. Track diagrams showing one possible choice of design for NOR, NOT, OR, and AND logic gates that use the interaction (blockage) between DNA walkers. Each walker stays on its own track, and all walkers begin stepping at the same time. Walkers denoted by **1** will always walk while **x** and **y** are walkers that will only start walking if their respective propositions are true. Walkers can block another track at the junctions marked with a red cross. The gate evaluates to **1** if the walker whose track has END at the end of it is indeed able to make it to the end without being blocked.

2 Directed Graph Abstraction

An input propositional formula, like those described in the previous section, can be used to find the topology for a localized DNA circuit. Every propositional variable is represented by a track, or linear array of DNA anchorages tethered to an origami tile, and a DNA walker. If the propositional variable holds the value **1**, then the walker will begin walking at the start of its track. Tracks that always take the value **1** have a DNA walker that will always start walking.

DNA walkers are able to perform universal Boolean logic if they are able to block other walkers on tracks that intersect their own. It is straightforward to construct a NOR gate, which is a functionally complete operator, using track blockage (Fig. 2, top left). Figure 2 shows both an AND gate constructed out of NOR gates and an alternative design that is simpler and uses fewer tracks.

A formula written in the propositional calculus, together with the chosen design for each gate, is enough to completely determine the topology of a corresponding DNA walker circuit. In the context of DNA localized computation systems, topology refers to "connectedness" of the tracks. Two tracks are said to be topologically connected if they intersect so that the walkers on both tracks can interact with each other.

More formally, for any chosen gate design, it is possible to describe the topology of each gate in terms of a directed graph $G_D = (V, E)$ where V is a set of vertices and E is a set of edges represented as ordered pairs. For the NOR gate,

$$V = \{1, x, y\}, \ E = \{(x, 1), (y, 1)\}.$$

The set V can be interpreted as the gate having three tracks, one for each of walkers $\{1, x, y\}$. The set of ordered pairs E indicates that the x walker blocks the 1 walker and the y walker blocks the 1 walker. Figure 2 shows one possible choice of gate design. Once a choice is made, directed graphs can be constructed for each gate as shown in Table 1. Directed graphs for the gates can be pieced together to form one directed graph for the whole circuit, capturing the topology of a DNA circuit that evaluates the propositional formula.

Table 1. The sets of edges E and vertices V in the directed graph that correspond to each logic gate. Subscripts are used to differentiate between unique tracks.

	V	E
NOR	$\{1, x, y\}$	$\{(x, 1), (y, 1)\}$
NOT	$\{1, x\}$	$\{(x, 1)\}$
OR	$\{1_1, 1_2, x, y\}$	$\{(1_2, 1_1), (x, 1_2), (y, 1_2)\}$
AND	$\{1, x, y\}$	$\{(1, x), (y, 1)\}$

Adding the directed graph abstraction between the propositional formula and its resulting track diagram has a number of advantages. At the topological level, the system becomes easier to analyze and simplify by automated means. The most immediate example is detecting certain redundancies, which can informally be thought of as double negatives. For any tracks a and b, if there exists a path $\{(a, 1_i), (1_i, 1_j), (1_j, b)\} \subset E$, then we may replace this path by $\{(a, b)\}$. In logical terms, this is the equivalent of writing $\neg(\neg b) = b$.

FANOUT

A key use of the directed graph structure is that it can represent circuits that cannot be posed in the propositional calculus. The most immediate example is FANOUT, which can be written as a directed graph:

$$V = \{x, 1_1, 1_2, 1_3, 1_4\},$$
$$E = \{(x, 1_1), (x, 1_2), (1_1, 1_3), (1_2, 1_4)\}.$$

As shown in Fig. 3, FANOUT requires an additional property of the walker-track system: The walker must be able to block a track and keep walking, blocking additional tracks thereafter. This property, as well as the FANOUT gate itself, is not necessary to perform universal Boolean logic. It can, however, be useful in simplifying track designs by using fewer walkers overall.

3   Localized DNA Circuits as Distributed Systems

This section shows how introducing another piece of information, a tolerance for the probability of error, allows one to use the circuit topology to find a circuit

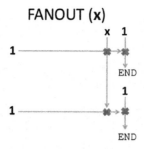

Fig. 3. A track diagram of a possible design choice for a FANOUT gate. This design requires a walker that can block a track and keep walking, blocking other tracks thereafter.

geometry. A geometry defines the length of each track and specifies the locations where the tracks intersect. The tools needed to find this information come from analyzing the localized DNA circuit as a distributed system.

A distributed system is a network of autonomous computers that can perform a coordinated action by passing messages between different computers in the network. When the anchorages on an origami tile are viewed as networked computers and the walker is viewed as the the message, a localized DNA computation system becomes a distributed system. The advantage of looking at the DNA circuit in this light is that it can be readily represented and analyzed as a Petri net.

3.1 Stochastic Petri Nets

The "burnt-bridges" walker-track system from [1] can be modeled as shown in Fig. 4, upper. The initial marking shows the DNA walker, represented by a token, on the first anchorage G1. The stochastic Petri net allows each walker a stepping rate, or the rate at which it steps forward onto the next anchorage. The transition from the first anchorage G1 to the second anchorage G2 fires at the rate at which the walker steps from one anchorage to the next. Such transitions require two tokens to fire. The tokens in the bottom row of nodes are used up as the walker steps forward, so another walker will not be able to step down the track after the current walker has finished. Physically, this bottom row of nodes represents the 5' end of the anchorages that are irreversibly nicked by the nicking enzyme. A reusable track can be represented in a similar fashion (Fig. 4, lower). In both cases, walkers are assumed to only step forward and remain on the last anchorage once they reach the end of their track.

Junctions between tracks are needed to implement the entire localized DNA circuit as a stochastic Petri net. Figure 5 shows a Petri net where a designated blocking walker (blue walker) can block another walker (green walker) if the blue walker arrives at the junction first. If the green walker arrives at the junction first, it steps through to the end of its track as normal.

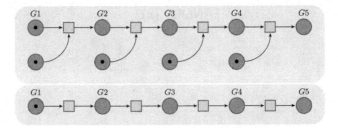

Fig. 4. *upper* Stochastic Petri net representation of the "burnt-bridges" walker from [1]. This track can only be used once, as the tokens in the bottom row of nodes are used up as the walker steps. *lower* Stochastic Petri net of a reusable track. Each transition only requires one token to fire, and no tokens are used up in the process.

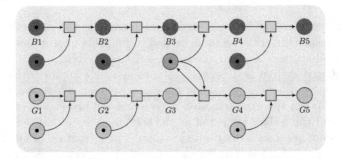

Fig. 5. Two tracks, green and blue, with a blocking junction on the third anchorage of each track (G3 and B3). If the blue walker arrives at the junction first, it can block the green track by using up the token of the shared node (shown in red).

A stochastic Petri net can be mapped directly onto a Markov process. If the DNA system can be posed as a Petri net, then it can be mapped onto a continuous time Markov chain (CTMC). Using this CTMC, the probability of certain properties of the system can be computed using techniques from formal verification.

3.2   Designing a System Using Formal Verification Techniques

The localized computation systems discussed thus far operate under the assumption that all DNA walkers, if they walk at all, begin walking at the same time and walk at the same rate. This imposes certain length restrictions on the tracks. In the NOT gate, for example, the track for the **1** walker must be sufficiently longer than the track for the x walker so that the **1** walker does not arrive at the junction first. If it does, a missed chance error has occurred because the x walker has missed its chance to block the **1** walker.

By representing the DNA system as a Markov chain and analyzing it for each possible combination of track lengths, one can search for a design that is optimal in the sense that it is compact and minimizes the probability of missed chance errors. Starting from the state corresponding to the initial marking of the Petri net, the system is allowed to evolve according to the CTMC. Earlier, it was assumed that all walkers can only step forward, if they step at all, and that walkers do not move forward once they are blocked or reach the end of their track. Hence, the Markov chain is an absorbing Markov chain, and an absorbing state will always be reached if the system runs to equilibrium. The absorbing states can be divided into two groups: one group that corresponds to a missed chance error for at least one junction, and another group that corresponds to correct operation. Analysis of the CTMC determines the probability with which the system ends up in missed chance error state or a correct state after it is allowed to run for a long time.

Prism provides a natural scripting language for representing complex systems as Markov chains in continuous time [5]. It also allows the user to evaluate the probability of certain conditions, such as the probability of eventually ending up in a certain state. For example, if a walker B is intended to block a walker G, Prism should check the following property:

$$P = ? \ [B=end \ \& \ G>intersection].$$

In Prism's language, this is the probability that the G walker has moved through the junction before the B walker has arrived to block it. This statement measures the probability of a missed chance error for that junction. Due to the modular nature of the Prism language, the code can be automatically generated from the directed graph structures described in the previous section.

Model checking can be used to determine the system with the shortest tracks that still has a missed chance error below a given tolerance. Finding this balance is critical: A compact system is easier to design and fit onto an origami tile, but tracks that are too short will cause missed chance errors. The output is the assignment of a natural number to each track for the smallest number of anchorages needed to stay within the specified tolerance for missed chance error. Assuming that a track is always blocked on its penultimate anchorage, this is sufficient to determine the track geometry of the system.

3.3 Illustrative Example 1: Track Design

This theoretical machinery forms the foundation for software that can design track systems. The only inputs required are a propositional formula, a choice of gate design, and a tolerance for missed chance error. The plot in Fig. 6 was automatically generated in this way.

The simplified propositional formula in Eq. 2 can be arranged into a parsing tree based on its logical connectives. Prism can be used to find the lengths (in anchorages) of each track that minimizes missed chance error. Using a tolerance of 0.15 probability for missed chance error results in the following optimized track lengths:

$$1_1 = 14 \text{ anchorages}, 1_2 = 1_3 = 7 \text{ anchorages}, x = y = z = 2 \text{ anchorages}.$$

Using the lengths of each individual track shown above and the blocking topology from the directed graph, a track design is generated that evaluates the original propositional formula (Fig. 6).

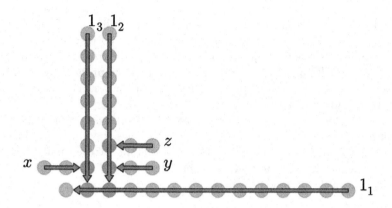

Fig. 6. Final design of the DNA localized computation system that can evaluate Eq. 2. Each anchorage is represented by a light orange circle. The individual tracks, along with their directionality, are indicated by orange arrows. Each track is labelled at its starting anchorage by the name of the walker that walks along it.

3.4   Illustrative Example 2: DNA Mechanism

Figures 7 and 8 show an example of a DNA walker mechanism that implements the junction in Fig. 5, whereby one walker (shown in blue) blocks another (shown in green). The notation used is similar to that of Microsoft Visual DSD. Single-stranded oligonucleotides are represented by arrows, as in Fig. 1. Domains are printed as short strings of characters, such as B, and a domain's reverse complement is appended with an asterisk. For example, the reverse complement of B is B^*. Certain domains contain a restriction site for a nicking enzyme. When a restriction site is completed by hybridization to its reverse complement, the location where the nicking enzyme will cut is indicated by a small red arrow. The mechanism requires that the nicking enzyme does not cut at certain domains that closely resemble the restriction site. These domains contain a restriction site mismatch, where one nucleotide in the restriction site has been altered. Domains containing a restriction site mismatch are indicated with an underscore. Locations where the anchorages are tethered to the origami tile are shown with an orange dot.

The green walker is made up of a short toehold domain G_t at the 3' end and a longer primary domain G_p at the 5' end. The blue walker is similar, with two important differences. First, the blue walker is in reverse orientation to the green walker. Its toehold domain is at the 5' end. Second, the blue walker has an extra G_t^* "tail" domain at the 3' end. A junction anchorage is located where the two tracks intersect and acts as a transducer, whereby the blue walker can convert

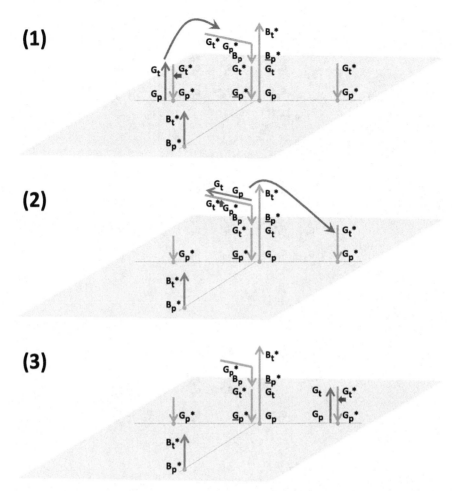

Fig. 7. The green walker arrives at the junction first. It can move through the junction and proceed down its track.

the junction anchorage from a normal anchorage to a trap that will block the green walker.

If the green track is unblocked (i.e. the blue walker has not arrived at the junction anchorage) then the green walker steps onto the auxiliary $[G_t^* \ G_p^* \ B_p]$ strand that is hybridized to the junction anchorage. The auxiliary strand resembles a normal green track anchorage (Fig. 7). The restriction site for the nicking enzyme is completed, the junction anchorage is cut, and the green walker steps on as normal. If the blue walker arrives at the junction anchorage first, it binds to the B_t^* toehold domain on the junction anchorage and displaces the auxiliary green anchorage from the junction (Fig. 8). This strand then diffuses away in solution. The 3' tail on the blue walker also displaces the G_t^* domain in the $[G_t^* \ \underline{G_p}^*]$ trapping strand. When the green walker arrives at the junction, it

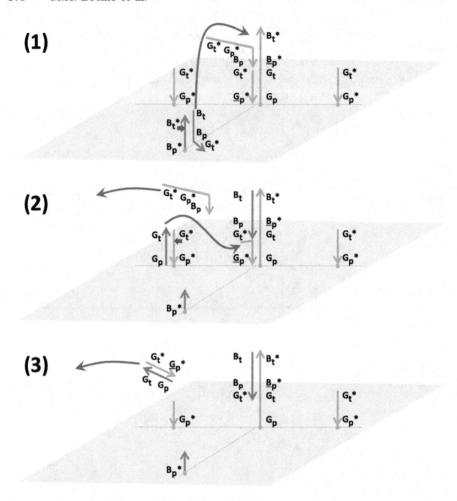

Fig. 8. The blue walker arrives at the junction first, exposing the toehold of a trap for the green walker. When the green walker arrives, it is trapped and the track is blocked.

binds to the toehold on the trapping strand that was exposed by the blue walker. The green walker hybridizes to the trapping strand and displaces it from the junction anchorage. A restriction site mismatch in the trapping strand makes the trap-green walker duplex inert and the duplex diffuses away in solution. Hence, the green walker has been blocked and removed from the track.

4   Conclusions

We have shown that localized DNA computation circuits can be analyzed as distributed systems. A propositional formula, a choice of gate design, and an error tolerance are enough to determine the geometry of the track system. There are simple blocking mechanisms for junction anchorages at the intersection between tracks that make such systems realizable.

The software and theory developed here can be improved upon by using it together with previously developed tools. One challenge in designing any DNA circuit is preventing unwanted interactions between domains. As the circuit gets more complex, these leaks become more and more difficult to identify by hand. Microsoft DSD already has automated means of detecting such leaks at the domain level. At the sequence level, NUPACK is an easy-to-use tool that can find suitable nucleotide sequences for a given design [12]. The three pieces of software can work together to first design a localized DNA circuit using the methods described above, then compile Microsoft DSD code to check for errors at the domain level, and finally compile NUPACK code to generate nucleotide sequences for each strand.

A challenge of working with DNA localized computation systems is making the system compact enough to fit on an origami tile while maintaining a low probability of missed chance error at the junctions. We can imagine a logic gate on an origami tile that, if it evaluates to **1**, activates a messenger strand that can set another walker stepping on a different origami tile. Such a mechanism can help alleviate the compactness issue. It would also increase the ability of localized DNA circuits to scale up and work with a higher reliability.

GitHub Repository

The software written to generate track designs is open source and freely available via the GitHub clone URL https://github.com/MBoemo/DLCC.git.

Acknowledgements. The authors would like to thank Andrew Phillips (Microsoft Research Cambridge), Marta Kwiatkowska (Oxford Computer Science), Alex Lucas (Oxford Physics), and Jonathan Bath (Oxford Physics) for guidance and helpful conversations.

References

1. Bath, J., Green, S.J., Turberfield, A.J.: A free-running DNA motor powered by a nicking enzyme. Angew. Chem. **117**, 4432–4435 (2005)
2. Benenson, Y., Gil, B., Ben-Dor, U., Adar, R., Shapiro, E.: An autonomous molecular computer for logical control of gene expression. Nature **429**, 423–429 (2004)
3. Dannenberg, F., Kwiatkowska, M., Thachuk, C., Turberfield, A.J.: DNA walker circuits: computational potential, design, and verification. In: Soloveichik, D., Yurke, B. (eds.) DNA 2013. LNCS, vol. 8141, pp. 31–45. Springer, Heidelberg (2013)
4. Joyner, D., Čertík, O., Meurer, A., Granger, B.E.: Open source computer algebra systems: SymPy. ACM Commun. Comput. Algebra **45**, 225–234 (2011)
5. Kwiatkowska, M., Norman, G., Parker, D.: PRISM 4.0: verification of probabilistic real-time systems. In: Gopalakrishnan, G., Qadeer, S. (eds.) CAV 2011. LNCS, vol. 6806, pp. 585–591. Springer, Heidelberg (2011)

6. Lakin, M.R., Youssef, S., Polo, F., Emmott, S., Phillips, A.: Visual DSD: a design and analysis tool for DNA strand displacement systems. Bioinformatics **27**, 3211–3213 (2011)

7. Lakin, M.R., Petersen, R., Gray, K.E., Phillips, A.: Abstract modelling of tethered DNA circuits. In: Murata, S., Kobayashi, S. (eds.) DNA 2014. LNCS, vol. 8727, pp. 132–147. Springer, Heidelberg (2014)

8. Phillips, A., Cardelli, L.: A programming language for composable DNA circuits. J. R. Soc. Interface **6**, S419–S436 (2009)

9. Rothemund, P.W.K.: Folding DNA to create nanoscale shapes and patterns. Nature **440**, 297–302 (2006)

10. Wickham, S.F.J., Bath, J., Katsuda, Y., Endo, M., Hidaka, K., Sugiyama, H., Turberfield, A.J.: A DNA-based molecular motor that can navigate a network of tracks. Nat. Nanotechnol. **7**, 169–173 (2012)

11. Yin, P., Turberfield, A.J., Sahu, S., Reif, J.H.: Design of an autonomous DNA nanomechanical device capable of universal computation and universal translational motion. In: Ferretti, C., Mauri, G., Zandron, C. (eds.) DNA 2004. LNCS, vol. 3384, pp. 426–444. Springer, Heidelberg (2005)

12. Zadeh, J.N., Steenberg, C.D., Bois, J.S., Wolfe, B.R., Pierce, M.B., Khan, A.R., Dirks, R.M., Pierce, N.A.: NUPACK: analysis and design of nucleic acid systems. J. Comput. Chem. **32**, 170–173 (2011)

On Low Energy Barrier Folding Pathways for Nucleic Acid Sequences

Leigh-Anne Mathieson and Anne Condon[✉]

Department of Computer Science, University of British Columbia,
Vancouver, BC, Canada
condon@cs.ubc.ca

Abstract. Secondary structure folding pathways correspond to the execution of DNA programs such as DNA strand displacement systems. It is helpful to understand the full diversity of features that such pathways can have, when designing novel folding pathways. In this work, we show that properties of folding pathways over a 2-base strand (a strand with either A and T, or C and G, but not all four bases) may be quite different than those over a 4-base alphabet. Our main result is that, for a simple energy model in which each base pair contributes -1, 2-base sequences of length n always have a folding pathway of length $O(n^3)$ with energy barrier at most 2. We provide an efficient algorithm for constructing such a pathway. In contrast, it is unknown whether minimum energy barrier pathways for 4-base sequences can be found efficiently, and such pathways can have barrier $\Theta(n)$. We also present several results that show how folding pathways with temporary and/or repeated base pairs can have lower energy barrier than pathways without such base pairs.

1 Introduction

Nucleic acid folding pathways—sequences of structures visited by DNA and RNA molecules as they fold—are interesting because they influence the shape and thus function of key agents of cellular processes [4]. Folding pathways are also very interesting to DNA nanotechnologists and molecular programmers because they are the realization of DNA programs for the creation of nano-materials, robots, logic circuits, artificial neural networks and much more [10,13,14,19,20].

Kinetics constrain nucleic acids to fold along pathways that tend to have low energy barriers. The energy barrier of a pathway, or simply the barrier, is the largest difference in free energy between any structure on the pathway and a subsequent structure. Specifically, if we are interested in folding pathways for a sequence s from an initial structure $\mathcal{I}$ to a final structure $\mathcal{F}$, where $\mathcal{I}$ has minimum free energy (MFE), then the energy barrier is the largest difference in free energy between $\mathcal{I}$ and any other structure along the pathway. We refer to a folding pathway from $\mathcal{I}$ to $\mathcal{F}$ with minimum barrier (taken over all possible folding pathways) as a min-barrier pathway. Several methods for computationally predicting nucleic acid folding pathways rely on energy barrier estimation [3,15].

© Springer International Publishing Switzerland 2015
A. Phillips and P. Yin (Eds.): DNA 2015, LNCS 9211, pp. 181–193, 2015.
DOI: 10.1007/978-3-319-21999-8_12

Moreover, designed nucleic acid systems such as DNA strand displacement systems ensure that the desired folding pathways have low energy barriers, while undesired alternatives have high barriers.

Thus there has been substantial work on methods for finding folding pathways between two given structures of a DNA or RNA strand s and in particular, finding min-barrier pathways (or approximations to these) [3,6]. These methods for computational prediction of folding pathways and energy barriers use reliable RNA or DNA thermodynamic and kinetic parameters [7], and mostly focus on pseudoknot free structures.

However, it can be helpful to work with simpler energy models, e.g., when the goal is to understand the computational complexity of folding pathway or energy barrier estimation, or to gain coarse-grained qualitative information on the shape of RNA folding landscapes [1,5,12]. Morgan and Higgs [8] studied how the energy barriers of min-barrier pathways of pseudoknot-free structures scale with strand length, assuming a simple energy model in which each base pair contributes -1 to the free energy of a structure. Their work considered so-called *direct* folding pathways in which the only base pairs that can be added along the folding pathway from structure $\mathcal{I}$ to structure $\mathcal{F}$ are those in $\mathcal{F} - \mathcal{I}$ and the only base pairs that can be removed are those in $\mathcal{I} - \mathcal{F}$. Thachuk et al. [17] showed that the *direct energy barrier problem (Direct-EBP)*, namely to determine whether there is a direct folding pathway from $\mathcal{I}$ to $\mathcal{F}$ with barrier of at most k, is NP-complete. Because of an earlier result of Thachuk et al. [16], the NP-completeness result holds whether or not the pathway can repeatedly remove and add back base pairs of $\mathcal{I}$ or $\mathcal{F}$ along the pathway.

The computational complexity of the more general energy barrier problem (EBP) remains open even for the simple energy model, where the EBP is to determine whether there is a possibly *indirect* pseudoknot-free folding pathway from $\mathcal{I}$ to $\mathcal{F}$ with barrier at most k. A pathway is indirect if so-called temporary and/or repeated base pairs can arise along the pathway, where a *temporary base pair* is one that is not in $\mathcal{I}$ or $\mathcal{F}$ but is in some other structure of the pathway, and a *repeated base pair* is one that is in some structure on the pathway (possibly the initial structure), then is removed and later added back again.

The main result of this paper is that there is indeed an efficient algorithm for the general energy barrier problem *for sequences over a 2-base alphabet*. For concreteness we state our result for sequences over the alphabet $\{A, U\}$, which we call AU-sequences. Our result shows that, for the simple energy model, not only is it possible to efficiently find a min-barrier pathway of length $O(|s|^3)$ from any initial MFE structure to any final MFE structure for any AU-sequence s, but that this pathway will have barrier 2 if the number of U's equals the number of A's and will have barrier 1 if the number of U's is not equal to the number of A's. In contrast, the minimum energy barrier of a sequence over a 4-base alphabet may be proportional to the length of the sequence.

Our algorithm relies heavily on the assumption of the simple energy model, but variants of the techniques involved, which are relatively straightforward and intuitive, may be useful also for more realistic energy models. The proof of our main result builds on the fact that the minimum free energy pseudoknot free structure of any AU-sequence s has energy $-q$, where q is the lesser of the number of A's in s and the number of U's in s.

Our main result raises two further questions that we address in this paper. First, our algorithm yields indirect barrier-1 or barrier-2 pathways, specifically, pathways with temporary base pairs. Dotu et al. [3] observed that there exist strands over $\{A, C, G, U\}$ whose min-barrier pathways are necessarily indirect. We strengthen Dotu et al.'s observation for the simple energy model, by showing that min-barrier pathways may also necessarily be indirect even for AU-sequences. Specifically we show that for any k, there is a length-$6k$ AU-sequence s, and minimum energy initial and final structures for s, such that any direct pathway from initial to final structure must have barrier at least $k + 1$, while there is a barrier-1 indirect pathway.

As noted above, it is not known whether there is an efficient (polynomial-time) algorithm for the EBP, for strands over $\{A,C,G,U\}$. It's conceivable that, because of the possibility that a min-barrier pathway must contain repeated base pairs, there exist infinitely many strands for which any min-barrier, indirect pathway from a given initial to a given target structure must have length that grows exponentially with the strand length. If this is the case, the EBP problem may be complete for PSPACE, a complexity class that is believed to include problems that are even harder than those in NP. Here we present the first example of a sequence s, initial structure $\mathcal{I}$ and final structure $\mathcal{F}$ such that the min-barrier pathway of pseudoknot-free structures has the property that base pairs which are in both $\mathcal{I}$ and $\mathcal{F}$ must be removed along the pathway, and then added back in again. This result for indirect pathways stands in contrast with the result of Thachuk et al. [16] that, for direct pathways, repeated base pairs are not necessary in min-barrier folding pathways.

The rest of this paper is organized as follows. Section 2 introduces notation and a preliminary result. We present our main result, namely our efficient algorithm for finding min-barrier folding pathways for AU-sequences, in Sect. 3. Our examples that illustrate why indirect pathways can have lower min-barrier than direct pathways for AU-sequences, and why pathways with repeats can have lower min-barrier than pathways without repeats, are in Sect. 4. Most of the proofs are omitted, because of space limitations. We present conclusions and directions for further work in Sect. 5.

2 Notation

Here we first introduce notation to describe nucleic acid secondary structure and folding pathways, and present a useful result on the free energy of minimum free energy structures. For an RNA sequence $s = s_1, s_2, \ldots, s_n$ (i.e., string over $\{A, C, G, U\}$), a *base pair* is an unordered pair $\{i, j\}$ where indices i and j are in the range $[1, \ldots, n]$, $i \neq j$, and the set of bases $\{s_i, s_j\}$ is either $\{A, U\}$ or $\{C, G\}$. (DNA is similar with T instead of U). A secondary structure $\mathcal{S}$ for s is a set of base pairs of s, such that no two intersect. Secondary structure is often represented as an arc diagram such as that in Fig. 1 (a), in which each base pair is represented as an arc that connects two bases of sequence s. For this reason, we often refer to a base pair as an arc, and refer to its indices as endpoints. We only consider pseudoknot-free structures: these are structures in which no arcs cross in the arc diagram representation. Equivalently, if a structure is pseudoknot free, then for all $\{i, j\}$ and $\{i', j'\}$ in the structure with $i < j$ and $i' < j'$, it is not the case that $i < i' < j < j'$ or $i' < i < j' < j$. Given a set S of arcs, a *narrowest* arc of S is an arc $\{i, j\}$ of S for which $|i - j|$ is minimal.

We use a simple energy model where each bond in a structure contributes -1 to the structure's free energy, and we denote the free energy of a structure $\mathcal{P}$ by $E(\mathcal{P})$. A *folding pathway*, π, from structure $\mathcal{I}$ to structure $\mathcal{F}$ is a sequence of pseudoknot-free secondary structures $\pi = \mathcal{P}_0, \mathcal{P}_1, ..., \mathcal{P}_m$ where $\mathcal{I} = \mathcal{P}_0$ and $\mathcal{F} = \mathcal{P}_m$. Each structure in the sequence differs from the structure directly before it by the addition or removal of exactly one base pair. When the first structure $\mathcal{I}$ on a folding pathway π is a MFE structure (as is always the case in this paper), the *energy barrier* of π is $\max_{1 \leq i \leq m} E(\mathcal{P}_0) - E(\mathcal{P}_i)$. Sometimes, rather than listing a given folding pathway, we list instead its *transformation sequence*, which is the sequence of arcs that are added or removed to obtain successive structures of the folding pathway. When listing the arcs of a transformation sequence, we use the prefices "+" and "−" to indicate whether the arc is added or removed. For example, if $\mathcal{I}$ is the structure $\{a_1, a_2, a_3\}$ with three arcs, then the transformation sequence $-a_1, +a_4, -a_2, +a_1$ corresponds to the folding pathway

$$\{a_1, a_2, a_3\}, \{a_2, a_3\}, \{a_2, a_3, a_4\}, \{a_3, a_4\}, \{a_1, a_3, a_4\}.$$

A U-index of s is a number u in the range $[1, ..., |s|]$ such that the base at position u of sequence s is U. An A-index of s is defined similarly, with A replacing U. If p is an arc then A-index(p) and U-index(p) denote the endpoints of p that are an A-index and a U-index, respectively. We say that an index i is *covered* by an arc p if i is in the range [A-index$(p) + 1$, U-index$(p) - 1$] if A-index$(p) <$ U-index(p) or the range [U-index$(p) + 1$, A-index$(p) - 1$] if U-index$(p) <$ A-index(p). Similarly, we say that an arc p is covered by arc p' if both endpoints of p are covered by p'.

An arc p' *separates* an index u from arc p if $p' \neq p$ and p' either covers u or covers p but does not cover both. Arc p' separates u from a set P of arcs if $p' \notin P$ and p' either covers u or all arcs in P, but not both.

For the simple energy model, the number of base pairs that could form in a secondary structure of an AU-sequence s is bounded by the minimum of the number of A's and the number of U's. Without loss of generality, suppose that s has at least as many U's as A's and let q be the number of A's. A simple stack-based algorithm can find a pseudoknot free structure with q base pairs in linear time:

Claim 1. *Let s be an AU-sequence with at least as many U's as A's, and let q be the number of A bases. There is a pseudoknot-free secondary structure S for s with q base pairs, and S can be generated in time $O(|s|)$.*

3   Low-Barrier Pathways for AU-Sequences

In this section we show how to find a folding pathway with barrier at most 2 from an initial MFE structure $\mathcal{I}$ to a final MFE structure $\mathcal{F}$ of an AU-sequence s. We consider two cases in the following two subsections: first, where the number of U's of s equals the number of A's and second, where there are more U's than A's. The case where there are more A's than U's can be handled in a manner symmetric to the case where there are more U's than A's and we do not discuss it further here.

3.1   AU-Sequences with an Equal Number of A's and U's

In the first case, a simple algorithm works to find a pathway with barrier 2, namely our FindBarrier2Pathway, Algorithm 1. This and later algorithms maintain a current structure S_{curr} which is initially set to $\mathcal{I}$; the algorithm repeatedly removes and adds arcs to S_{curr} until the structure $\mathcal{F}$ is reached, and the resulting sequence of structures forms the folding pathway. In this case, the algorithm adds the arcs of $\mathcal{F}$ to S_{curr} in narrowest-first order. Before adding arc f_{nar}, the two arcs of S_{curr} that share an endpoint with f_{nar} must first be removed; then f_{nar} and one additional arc are added in order to avoid a barrier of more than 2. At the start of each iteration, $\mathcal{F}_{frozen}$ is the set of arcs of $\mathcal{F}$ that have already been added to S_{curr} (these arcs are "frozen" in the sense that they will not be subsequently removed from S_{curr}). Claim 2 asserts that this can be done without introducing pseudoknots. We also note that if the number of U's is not equal to the number of A's, Algorithm 1 is not correct.

Claim 2. *The pathway π produced by FindBarrier2Pathway (Algorithm 1) on input s, $\mathcal{I}$, $\mathcal{F}$ is a valid barrier-2 pathway from $\mathcal{I}$ to $\mathcal{F}$ where no structure in the pathway contains pseudoknots. The pathway produced has length at most 4 times the number of arcs in an MFE structure of s.*

3.2   AU-Sequences with More U's Than A's

If sequence s has more U's than A's, there is a barrier-1 pathway from MFE structure $\mathcal{I}$ to MFE structure $\mathcal{F}$. Here we present our FindPathway algorithm, Algorithm 2, which constructs this pathway.

Starting with a current structure S_{curr} that is set to the initial structure $\mathcal{I}$, FindPathway repeatedly selects an arc f of $\mathcal{F}$ that is not in the current structure. For each f, it calls the ResolveConflicts algorithm, Algorithm 3, which updates the current structure via a barrier-1 pathway that removes any arcs that conflict with, i.e., form a pseudoknot with, f, while also ensuring that arcs of $\mathcal{F}$ that were added in earlier iterations—so-called *frozen* arcs—are not removed. Once ResolveConflicts is done, the FindPathways algorithm adds f to S_{curr} and arc f is also frozen. As we show later, the order in which the arcs of $\mathcal{F}$ are added by FindPathway ensures that ResolveConflicts can proceed within barrier 1.

Algorithm 1. Find a barrier-2 pathway for an AU-sequence with #U's = #A's

procedure FindBarrier2Pathway $(s, \mathcal{I}, \mathcal{F})$

  Input:

    a sequence $s \in \{A,U\}^*$, with an equal number of U's and A's

    an initial MFE structure $\mathcal{I}$ for s

    a final MFE structure $\mathcal{F}$ for s

  Output:

    a valid pathway π from $\mathcal{I}$ to $\mathcal{F}$ with barrier 2

  $\mathcal{S}_{curr} = \mathcal{I}$; $\pi \leftarrow$ empty pathway; $\mathcal{F}_{frozen} \leftarrow \emptyset$

  while $\mathcal{F}_{frozen} \neq \mathcal{F}$ **do**

    $f_{nar} \leftarrow$ a narrowest arc such that $f_{nar} \in \mathcal{F}$ but $f_{nar} \notin \mathcal{F}_{frozen}$

    if $f_{nar} \notin \mathcal{S}_{curr}$ **then**

      $f_a \leftarrow$ the arc of $\mathcal{S}_{curr}$ with endpoint A-index(f_{nar})

      $f_u \leftarrow$ the arc of $\mathcal{S}_{curr}$ with endpoint U-index(f_{nar})

      $p \leftarrow$ the arc with endpoint U-index(f_a) and A-index(f_u)

      remove f_a from $\mathcal{S}_{curr}$; $\pi \leftarrow \pi, \mathcal{S}_{curr}$

      remove f_u from $\mathcal{S}_{curr}$; $\pi \leftarrow \pi, \mathcal{S}_{curr}$

      add p to $\mathcal{S}_{curr}$; $\pi \leftarrow \pi, \mathcal{S}_{curr}$

      add f_{nar} to $\mathcal{S}_{curr}$; $\pi \leftarrow \pi, \mathcal{S}_{curr}$

    end if

    add f_{nar} to $\mathcal{F}_{frozen}$

  end while

  return π

We next describe the ResolveConflicts algorithm, while also introducing definitions that are used in the algorithm descriptions. These definitions are with respect to the inputs to ResolveConflicts, namely a "current" pseudoknot-free secondary structure $\mathcal{S}_{curr}$ for s, a subset $\mathcal{F}_{frozen}$—the frozen arcs of $\mathcal{S}_{curr}$, and an additional arc f of $\mathcal{F}$ that is not yet in $\mathcal{S}_{curr}$. Let conflict(f) be the set of arcs of $\mathcal{S}_{curr}$ that form a pseudoknot with f, except that the arc of $\mathcal{S}_{curr}$ from the A-index endpoint of f is excluded.

ResolveConflicts repeatedly removes the arcs of conflict(f), keeping the barrier low by "repairing" the A-indices of these conflicting arcs with other available U-indices. To do this, ResolveConflicts first identifies a set $\mathcal{U}$ of currently unpaired U-indices that can *indirectly repair* conflict(f). A U-index u of s can indirectly repair conflict(f) if u is unpaired in $\mathcal{S}_{curr}$ and no arc of $\mathcal{F}_{frozen} \cup \{f\}$ separates an index of $\mathcal{U}$ from conflict(f). (If an arc p separates u from some arc of conflict(f) then p must separate u from all arcs of conflict(f).) It is the case (details omitted) that conflict(f) is indeed *repairable*, that is, there is a set $\mathcal{U}$ of $|$conflict$(f)|$ U-indices that can indirectly repair conflict(f). However, it may not be possible for ResolveConflicts to simply remove an arc p from conflict(f) and pair its A-index with a U-index of $\mathcal{U}$ without creating a pseudoknot. We say that an unpaired U-index u *can directly repair* an arc p if no arc of $\mathcal{S}_{curr} \cup \{f\} - \{p\}$ separates u from A-index(p). The inner while loop of ResolveConflicts finds a pathway that can "convert" an unpaired base u of $\mathcal{U}$ into an unpaired base that can directly repair an arc p of conflict(f). The outer loop of ResolveConflicts then removes p and adds (A-index(p), u) to $\mathcal{S}_{curr}$, thereby reducing the number of arcs that conflict with f. ResolveConflicts ends once all conflicts are removed.

Claims 3 and 4 assert that the ResolveConflicts and FindPathway algorithms are correct, leading to our main result of this section, Theorem 1.

Claim 3. *ResolveConflicts is correct, that is, produces an output with the properties specified at the top of the algorithm description, given an input with the properties specified at the top of the algorithm description.*

Claim 4. *FindPathway is correct.*

Theorem 1. *Let $(s, \mathcal{I}, \mathcal{F})$ be an AU-instance of the EBP. A barrier-2 pathway of length $O(|s|)$ can be found in $O(|s|)$ steps for (s, I, F). Moreover, if the number*

Algorithm 2. Find a valid barrier-1 folding pathway from initial structure $\mathcal{I}$ to final structure $\mathcal{F}$, for a sequence s that has more U's than A's.

---

algorithm FindPathway($\mathcal{I}, \mathcal{F}, s$)

Input:
   a sequence $s \in \{A, U\}^*$, with more U's than A's
   an initial MFE pseudoknot-free structure $\mathcal{I}$ for s
   a final MFE pseudoknot-free structure $\mathcal{F}$ for s
Output:
   a valid pseudoknot-free folding pathway π from $\mathcal{I}$ to $\mathcal{F}$ with barrier 1

$\mathcal{S}_{curr} = \mathcal{I}$; $\pi \leftarrow$ empty pathway; $\mathcal{F}_{frozen} \leftarrow \emptyset$
if in $\mathcal{F}$, some U-index is unpaired and not covered by any arc **then**
   let U-chosen be any such U-index
else
   let U-chosen be any U-index that is unpaired in $\mathcal{F}$ and is covered
   by a narrowest arc of $\mathcal{F}$ (among those arcs covering unpaired U-indices)
end if

while some arc of $\mathcal{F} - \mathcal{F}_{frozen}$ does not cover U-chosen **do**
   let f be a narrowest such arc in $\mathcal{F} - \mathcal{F}_{frozen}$
   $(\mathcal{S}', \pi') \leftarrow$ ResolveConflicts($s, \mathcal{S}_{curr}, \mathcal{F}_{frozen}, f$)
   append π' to π; $\mathcal{S}_{curr} \leftarrow \mathcal{S}'$
   remove the arc of $\mathcal{S}_{curr}$ containing A-index(f) as an endpoint; $\pi \leftarrow \pi, \mathcal{S}_{curr}$
   add f to $\mathcal{S}_{curr}$; $\pi \leftarrow \pi, \mathcal{S}_{curr}$; $\mathcal{F}_{frozen} \leftarrow \mathcal{F}_{frozen} \cup \{f\}$
end while// all arcs of $\mathcal{F} - \mathcal{F}_{frozen}$ cover U-chosen

while $\mathcal{S}_{curr} \neq \mathcal{F}$ **do**
   let f be the widest arc in $\mathcal{F} - \mathcal{F}_{frozen}$
   $(\mathcal{S}', \pi') \leftarrow$ ResolveConflicts($s, \mathcal{S}_{curr}, \mathcal{F}_{frozen}, f$)
   append π' to π; $\mathcal{S}_{curr} \leftarrow \mathcal{S}'$
   remove the arc of $\mathcal{S}_{curr}$ containing A-index(f) as an endpoint; $\pi \leftarrow \pi, \mathcal{S}_{curr}$
   add f to $\mathcal{S}_{curr}$; $\pi \leftarrow \pi, \mathcal{S}_{curr}$; $\mathcal{F}_{frozen} \leftarrow \mathcal{F}_{frozen} \cup \{f\}$
end while
return π

Procedure 3. Find a valid barrier-1 pathway from an input MFE structure $\mathcal{S}_{curr}$ for sequence s to an updated MFE structure $\mathcal{S}_{curr}$ for s, where the updated $\mathcal{S}_{curr}$ contains all arcs in $\mathcal{F}_{frozen}$, a subset of $\mathcal{S}$, and such that conflict(f) is empty.

---

procedure ResolveConflicts (s,$\mathcal{S}_{curr}$,$\mathcal{F}_{frozen}$,f)

   Input:
     sequence $s \in \{A,U\}^*$, with more U's than A's, MFE structure $\mathcal{S}_{curr}$ for s,
     $\mathcal{F}_{frozen} \subset \mathcal{S}_{curr}$ and arc $f \notin \mathcal{F}_{frozen}$ such that conflict(f) is repairable
   Output:
     updated MFE structure $\mathcal{S}_{curr}$ for s such that $\mathcal{F}_{frozen} \subseteq \mathcal{S}_{curr}$ and conflict(f) is
     empty a valid barrier-1 pathway π' from the input $\mathcal{S}_{curr}$ to the output $\mathcal{S}_{curr}$

   $\pi' \leftarrow$ empty pathway
   let $\mathcal{U}$ be a set of $|\,$conflict(f)$|$ U-indices that can indirectly repair conflict(f)
   while $|\,$conflict(f)$| > 0$ **do**
     // create a U-index that can directly repair some arc of conflict(f)
     select some u in $\mathcal{U}$ and remove u from $\mathcal{U}$
     while u cannot directly repair any arc of conflict(f) **do**
       let p be an arc that separates u from conflict(f), such that u can directly repair
       p
       remove p from $\mathcal{S}_{curr}$; $\pi \leftarrow \pi, \mathcal{S}_{curr}$
       add $\{A\text{-index}(p), u\}$ to $\mathcal{S}_{curr}$; $\pi \leftarrow \pi, \mathcal{S}_{curr}$
       $u \leftarrow$ U-index(p)
     end while
     choose arc $p \in$ conflict(f) such that u can directly repair p
     remove p from $\mathcal{S}_{curr}$; $\pi \leftarrow \pi, \mathcal{S}_{curr}$
     add $\{A\text{-index}(p), u\}$ to $\mathcal{S}_{curr}$; $\pi \leftarrow \pi, \mathcal{S}_{curr}$
   end while
   return $(\mathcal{S}_{curr}, \pi')$

---

of A's of s does not equal the number of U's of s, a barrier-1 pathway of length $O(|s|^3)$ can be found in $O(|s|^3)$ time.

Proof. Claim 2 shows that Algorithm 1, FindBarrier2Pathway, finds a barrier-2, length $O(n)$ pathway for an AU-instance $(s, \mathcal{I}, \mathcal{F})$ of the EBP. The number of steps of the algorithm is $O(|s|)$, since there are $\mathcal{F} \leq |s|$ iterations of the whle loop, each taking $O(1)$ steps.

    Claim 4 shows that FindPathway, Algorithm 2 finds a barrier-1 pathway when the AU-instance is such that the number of U's is greater than the number of A's (and by swapping U's and A's in the algorithm works when the number of A's is greater than the number of U's). To bound the length of the pathway, we first need to bound the number of steps in ResolveConflicts, Algorithm 3 (which is called by FindPathway). Each iteration of the inner while loop of ResolveConflicts reduces the number of arcs that separate u from conflict(f) by 1, and thus the number of iterations is $O(|s|)$. Each iteration has $O(1)$ steps and thus the total number of steps per iteration of the inner while loop, and the length of the pathway segment generated, is $O(|s|)$. Each iteration of the outer while loop reduces the size of conflict(f) by 1, using $O(1)$ steps beyond those of the inner while loop. Therefore, the total number of steps of ResolveConflicts is $O(|s|^2)$, and the total length of the pathway segment generated is also $O(|s|^2)$. For each arc of $\mathcal{F}$ that is added to $\mathcal{F}_{frozen}$, FindPathway calls ResolveConflicts once, and takes $O(1)$ additional steps. Thus, the overall length of the pathway generated by FindPathway is $O(|s|^3)$, and this also bounds the total number of steps taken by the algorithm (including calls to ResolveConflicts).

4   On Min-Barrier Pathways that Are Necessarily Indirect Pathways or Contain Repeat Base Pairs

Theorem 2. *For any k, there is a length-6k AU-sequence with minimum energy initial and final structures such that any direct pathway from initial to final structure must have barrier at least $k + 1$, while there is a barrier-1 indirect pathway.*

Proof. The length-6k AU-sequence is $A^k U^k U^k A^k U^k U^k$, where here X^k is the letter X repeated k times. The initial and final structures are $\mathcal{I} = \binom{k \ . \ k}{}^k \binom{k \ . \ k}{}^k$ and $\mathcal{F} = \binom{k}{}\binom{k \ . \ k}{}^k \big)^k \ . \ k$ respectively. That is, $\mathcal{I}$ has two disjoint hairpin-forming stems that we refer to as the left and right stems, while $\mathcal{F}$ has one stem nested in another; we refer to these as the inner and outer stems. Note also that the set of A-indices of $\mathcal{I}$'s left stem equals the set of A-indices of $\mathcal{F}$'s outer stem, and the set of A-indices of $\mathcal{I}$'s right stem equals the set of A-indices of $\mathcal{F}$'s inner stem. Figure 1 illustrates the sequence and initial and final structures for $k = 3$.

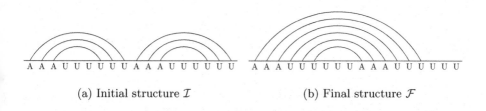

(a) Initial structure $\mathcal{I}$ (b) Final structure $\mathcal{F}$

Fig. 1. Illustration of the construction of Theorem 2 for $k = 3$.

We first show that any direct pathway must have barrier at least $k + 1$. Let $P = p_1, p_2, \ldots, p_{|P|}$ be a direct pathway from $\mathcal{I}$ to $\mathcal{F}$. Let a be the first arc of $\mathcal{F}$ that appears in a structure of pathway P, say structure p_i. By definition of a direct pathway, the only arcs that can be in p_{i-1} are either arcs from $\mathcal{I}$ or $\mathcal{F}$. However, since a is the first arc of $\mathcal{F}$ to appear in a structure of P, with a appearing first in p_i, p_{i-1} contains no arc of $\mathcal{F}$. If a is in the outer stem of $\mathcal{F}$, then p_{i-1} also contains none of the k arcs from the right stem of $\mathcal{I}$; otherwise such arcs would cause a pseudoknot with a in p_i. Furthermore, at least one arc from the left stem of $\mathcal{I}$, namely the arc that shares an endpoint with a, is not in

p_{i-1}. Therefore at most $k-1$ arcs of P are in p_{i-1}; since $\mathcal{I}$ and $\mathcal{F}$ have $2k$ arcs, p_{i-1} causes the barrier of the path to be $k+1$. A similar argument shows that if a is in the inner stem of $\mathcal{F}$, then p_{i-1} also contains at most $k-1$ arcs and thus the barrier is $k+1$.

Next we show that there is an indirect, barrier-1 pathway from $\mathcal{I}$ to $\mathcal{F}$. The pathway has several stages. First, the right stem of $\mathcal{I}$ is replaced by a narrower stem to obtain the structure $(^{k}.^{k})^{k}(^{k}_{)}^{k}.^{k}$. This can be done via a barrier-1 pathway in which the arcs of $\mathcal{I}$'s right stem are replaced by narrower arcs, in narrowest-first order. Then, the left stem of $\mathcal{I}$ can be replaced by a stem that spans from the leftmost A's to the rightmost U's of the sequence, via a barrier-1 pathway, thereby reaching current structure $(^{k}.^{k}.^{k}(^{k})^{k})^{k}$ Then replace the inner stem of the current structure with the inner stem of $\mathcal{F}$. Finally, replace the wide stem of the current structure with the outer stem of $\mathcal{F}$.

Theorem 3. *There exists an AU-sequence s, with corresponding initial structure $\mathcal{I}$ and final structure $\mathcal{F}$ where there is an indirect pathway with repeats with a lower energy barrier than the energy barrier than that of any direct pathway.*

Proof. Consider the sequence and structures $\mathcal{I}$ and $\mathcal{F}$ of Fig. 2.

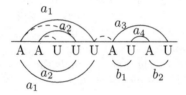

Fig. 2. An initial structure $\mathcal{I} = \{a_1, a_2, a_3, a_4\}$ (top) and a final structure $\mathcal{F} = \{a_1, a_2, b_1, b_2\}$ (bottom) for sequence AAUUUAUAU, such that there is no barrier-1 pathway without repeats from $\mathcal{I}$ to $\mathcal{F}$. Additional dashed arcs are required for a barrier-1 pathway.

We first consider possible barrier-1 pathways without repeats from structure $\mathcal{I}$. Note that since a_1 and a_2 are in $\mathcal{F}$ that in any pathway without repeats they cannot be removed as re-adding either of them would cause a repeat. So we move onto adding b_1 and b_2 without introducing a repeat, and to add either requires first removing both a_3 and a_4, which means that any pathway that does not allow repeats must be barrier-2.

So, we are left to demonstrate that there exists a barrier-1 pathway from $\mathcal{I}$ that contains repeats. We will need to add the dashed arcs in Fig. 2, so of the two nested dashed arcs, let's denote the narrower one by t_1 and the wider one by t_2, and the remaining dashed arc shall be t_3.

The following transformation sequence is barrier-1, and requires a_1 and a_2 to repeat; as an arc is added immediately after every arc that is removed, we have a barrier-1 pathway.

$$\mathcal{I} = -a_2, +t_1, -a_1, +t_2, -a_3, +t_3, -a_4, +b_2, -t_3, +b_1, -t_2, +a_1, -t_1, +a_2$$

5   Conclusions and Future Work

In this paper, for sequences over two bases, we show how to efficiently find min-barrier, pseudoknot-free pathways from initial to final MFE structures, for an energy model that assigns "-1" to each base pair (Theorem 1). In contrast, the computational complexity of finding such min-barrier pathways for sequences over four bases is unknown, and the problem may well be computationally intractable. We also show that min-barrier pathways for sequences over two bases may necessarily be indirect, i.e., involve base pairs that are neither in the initial nor final structures, and that direct pathways for such sequences may have a minimum energy barrier that is proportional to the length of the sequence (Theorem 2). Thirdly, we show that a weak form of arc repetition may be necessary in a min-barrier pathway (Theorem 3).

There are several ways in which our results could be improved. Our algorithm yields a $O(n^3)$ bound on the length of a barrier-1 pathway between two MFE structures of a length-n AU-sequence. We expect that this can be reduced, by carefully choosing the order in which u's are chosen from $\mathcal{U}$ in the while loop of the ResolveConflicts algorithm, the order in which conflicts are repaired, and perhaps also the order in which arcs are added to $\mathcal{F}_{frozen}$. Can the pathway length be reduced to $O(n)$? Another question is whether the problem of finding min-barrier, direct, pseudoknot-free pathways has an efficient algorithm (recall that for 4-base sequences, the problem is NP-hard [17]).

A significant limitation of our results is that the simple energy model ignores critical aspects of real RNA thermodynamics, such as base stacking energies, the energy costs of helix formation and loops, and the fact that hairpin loops have at least three unpaired nucleotides between their innermost paired bases. Another concern is that the model ignores pseudoknots, particularly given that pseudoknots may occur in intermediate structures along a folding pathway to a native structure, even if there is no pseudoknot in the native structure [18]. A first step forward in improving the model would be to have an energy of "-1" per stacked pair. It may be feasible to provide proofs as to whether, for this model, the energy barrier for sequences over two bases is bounded. NP-hardness of the energy barrier problem for the stacked pair model, for either two-base or four-base sequences, would suggest that molecular programs could perhaps be encoded within a DNA or RNA strand; the program could be executed via the strand's folding pathway, with the number of steps being exponential in the strand length. Alternatively, an efficient algorithm might indicate limits to the potential for long computations with a single nucleic acid strand, but could be useful in practice for finding folding pathways.

Given that it will likely be difficult to prove rigorous results for more realistic energy models, empirical computational studies could be very useful in elucidating whether the contrasting properties of two-base and four-base folding pathways described in this paper reflect the properties of two-base versus four-base sequences with respect to realistic energy models. The following questions could fruitfully be investigated empirically. Are there significant differences in min-energy barriers of pathways between low-energy structures of random versus biological sequences? Of two-base and four-base sequences? In particular, is the the min-energy barrier of any two-base sequence bounded by a constant that is independent of the sequence length? Are two-base sequences more likely to quickly fold to their MFE structures, compared with four-base sequences? Or alternatively, is it possible to design a two-base sequence with a kinetic trap that causes the sequence to fold slowly to its MFE state? Insights on questions such as these may be relevant to a hypothesis that in the early history of life, a precursor to RNA contained only two nucleotides [2,9]. Are there examples of biological molecules that follow indirect folding pathways, or which repeatedly add and remove base pairs or stems (rather than following shorter, possibly higher-barrier pathways)? We plan to study these questions using available software tools for folding pathway and energy barrier prediction.

Acknowledgments. We thank the reviewers of the paper for their detailed and thoughtful comments, for raising their significant concerns about the value of our results in light of the underlying simplistic energy model, and for pointing us to the work of Reader and Joyce [11]. Their comments on follow-on work are reflected in Sect. 5 and, while beyond the scope of what we could address in our revisions, will guide us in our future work.

References

1. Clote, P.: An efficient algorithm to compute the landscape of locally optimal RNA seconary structures with respect to the nussinov-jacobson enery model. J. Comput. Biol. **12**, 83–101 (2005)
2. Crick, F.H.C.: The origin of the genetic code. J. Mol. Biol. **38**, 367–379 (1968)
3. Dotu, I., Lorenz, W.A., Van Hentenryck, P., Clote, P.: Computing folding pathways between RNA secondary structures. Nucleic Acids Res. **38**(5), 1711–1722 (2010)
4. Flamm, C., Fontana, W., Hofacker, I.L., Schuster, P.: RNA folding at elementary step resolution. RNA **6**(3), 325–338 (2000)
5. Flamm, C., Hofacker, I.L., Stadler, P.F., Wolfinger, M.T.: Barrier trees of degenerate landscapes. Zeitschrift für physikalische chemie **216**, 155–174 (2002)
6. Geis, M., Flamm, C., Wolfinger, M.T., Tanzer, A., Hofacker, I.L., Middendorf, M., Mandl, C., Stadler, P.F., Thurner, C.: Folding kinetics of large RNAs. J. Mol. Biol. **379**, 160–173 (2008)
7. Mathews, D.H., Sabina, J., Zuker, M., Turner, D.H.: Expanded sequence dependence of thermodynamic parameters improves prediction of RNA secondary structure. J. Mol. Biol. **288**, 911940 (1999)
8. Morgan, S.R., Higgs, P.G.: Barrier heights between ground states in a model of RNA secondary st ructure. J. Phys. A: Math. Gen. **31**, 3153–3170 (1998)
9. Orgel, L.E.: Evolution of the genetic apparatus. J. Mol. Biol. **38**, 381–393 (1968)
10. Qian, L., Winfree, E., Bruck, J.: Neural network computation with dna strand displacement cascades. Nature **475**, 368372 (2011)
11. Reader, J.S., Joyce, G.F.: A ribozyme composed of only two different nucleotides. Nature **420**, 841844 (2002)
12. Schuster, P., Fontana, W., Stadler, P., Hofacker, I.L.: From sequences to shapes and back: a case study in RNA secondary structures. In: Proceedings-Royal Society of London, Biological sciences, pp. 279–284 (1994)

13. Seelig, G., Soloveichik, D., Zhang, D.Y., Winfree, E.: Enzyme-free nucleic acid logic circuits. Science **314**, 1585–1588 (2006)
14. Simmel, F.C., Dittmer, W.U.: DNA nanodevices. Small **1**(3), 284–299 (2005)
15. Tang, X., Thomas, S., Tapia, L., Giedroc, D.P., Amato, N.M.: Simulating RNA folding kinetics on approximated energy landscapes. J. Mol. Biol. **381**, 1055–1067 (2008)
16. Thachuk, C., Manuch, J., Rafiey, A., Mathieson, L.A., Stacho, L., Condon, A.: An algorithm for the energy barrier problem without pseudoknots and temporary arcs. In: Proceedings of the Pacific Symposium on Biocomputing (2010)
17. Thachuk, C., Manuch, J., Stacho, L., Condon, A.: NP-completeness of the direct energy barrier height problem without pseudoknots. Natural Comput. **10**(1), 391–405 (2011)
18. Wiebe, N.J.P., Meyer, I.M.: Transat a method for detecting the conserved helices of functional rna structures, including transient, pseudo-knotted and alternative structures. PLoS Comput. Biol. **6**(6), e1000823 (2010)
19. Yin, P., Choi, H.M.T., Calvert, C.R., Pierce, N.A.: Programming biomolecular self-assembly pathways. Nature **451**, 318–322 (2008)
20. Yurke, B., Turberfield, A.J., Mills Jr., A.P., Simmel, F.C., Neumann, J.L.: A DNA-fuelled molecular machine made of DNA. Nature **406**, 605–608 (2000)

Stochastic Simulation of the Kinetics of Multiple Interacting Nucleic Acid Strands

Joseph Malcolm Schaeffer[1,2], Chris Thachuk[1], and Erik Winfree[1(✉)]

[1] California Institute of Technology, Pasadena, USA
winfree@caltech.edu
[2] Autodesk Research, San Francisco, USA

Abstract. DNA nanotechnology is an emerging field which utilizes the unique structural properties of nucleic acids in order to build nanoscale devices, such as logic gates, motors, walkers, and algorithmic structures. Predicting the structure and interactions of a DNA device requires effective modeling of both the thermodynamics and the kinetics of the DNA strands within the system. The kinetics of a set of DNA strands can be modeled as a continuous time Markov process through the state space of all secondary structures. The primary means of exploring the kinetics of a DNA system is by simulating trajectories through the state space and aggregating data over many such trajectories. We expand on previous work by extending the thermodynamics and kinetics models to handle multiple strands in a fixed volume, in a way that is consistent with previous models. We developed data structures and algorithms that allow us to take advantage of local properties of secondary structure, improving the efficiency of the simulator so that we can handle reasonably large systems. Finally, we illustrate the simulator's analysis methods on a simple case study.

1 Introduction

Dynamic DNA nanotechnology [29] is an emerging field that utilizes the unique structural properties of nucleic acids in order to build nanoscale devices, such as conformational motors [27], hybridization catalysts [21], logic gates [14,19], analog circuits [2,26,30], triggered self-assembly [6,26], polymerization motors [22], molecular walkers [13,20], and molecular robots [9,12] that operate even in the absence of enzymes and other sophisticated non-nucleic-acid chemistry. These devices are built out of DNA strands whose sequences have been carefully designed in order to control their secondary structure—the hydrogen bonding state of the bases within the strand (called "base-pairing"). This base-pairing is used to not only control the physical structure of the device, but also to enable specific interactions between different components of the system, such as allowing, for example, a DNA strand that catalytically triggers the assembly of two components. Predicting the structure and interactions of a DNA device requires effective modeling of both the thermodynamics and the kinetics of the DNA strands within the system. Thermodynamic models can be used to make

© Springer International Publishing Switzerland 2015
A. Phillips and P. Yin (Eds.): DNA 2015, LNCS 9211, pp. 194–211, 2015.
DOI: 10.1007/978-3-319-21999-8_13

equilibrium predictions for these systems, allowing us to look at questions like "Is the assembled end-product a well-formed molecular structure, and is it energetically favorable?", while kinetics models allow us to predict the non-equilibrium dynamics, such as "How quickly will the catalytic pathway take place?" Although the thermodynamics of multiple interacting DNA strands is a well-studied model [3,4], which allows for both analysis and design of DNA devices [5,28], previous work on secondary structure kinetics models only explored the kinetics of how a single strand folds on itself [7,25].

The kinetics of a set of DNA strands can be modeled as a continuous time Markov process through the state space of all secondary structures. Due to the exponential size of this state space it is computationally intractable to obtain an analytic solution for most problem sizes of interest. Thus the primary means of exploring the kinetics of a DNA system is by simulating trajectories through the state space and aggregating data over many such trajectories. We present here the **Multistrand** kinetics simulator, which extends previous work [7] by using the multiple strand thermodynamics model [4] (a core component for calculating transition rates in the kinetics model), adding new terms to the thermodynamics model to account for stochastic modeling considerations, and by adding new kinetic moves that allow bimolecular interactions between strands. In Ref. [18], we prove that this new kinetics and thermodynamics model is consistent with the prior work on multiple strand thermodynamics models [4].

The Multistrand simulator is based on the Gillespie algorithm [8] for generating statistically correct trajectories of a stochastic Markov process. We developed data structures and algorithms that take advantage of local properties of secondary structures. These algorithms enable the efficient reuse of the basic objects that form the system, such that only a very small part of the state's neighborhood information needs to be recalculated with every step. A key addition was the implementation of algorithms to handle the new kinetic steps that occur between different DNA strands, without increasing the time complexity of the overall simulation. These improvements lead to a reduction in worst case time complexity of a single step and also lead to additional improvements in the average case time complexity.

What data does the simulation produce? At the very simplest, the simulation produces a full kinetic trajectory through the state space—the exact states it passed through, and the time at which it reached them. A small system might produce trajectories that pass through hundreds of thousands of states, and that number increases rapidly as the system gets larger. Going back to our original question, the type of information a researcher hopes to get out of the data could be very simple: "How quickly will the catalytic pathway take place?", with the implied question of whether it's worth it to actually purchase the particular DNA strands composing the catalyst system and perform an experiment, or go back to the drawing board and redesign the device. One way to acquire that type of information is to look at the first time in the trajectory where we reached the "assembly has been catalyzed" state, and record that information for a large number of simulated trajectories in order to obtain a useful answer. We designed and implemented new simulation modes that allow the full trajectory data to be condensed during generation into only the pieces the user cares about for their particular question. This analysis tool also required the development of flexible ways to talk about states that occur in trajectory data; if someone wants data on when or how the catalyst acted, we have to be able to express that in terms of the Markov process states which meet that condition.

2   The Model

2.1   System Specification

We are interested in simulating nucleic acid molecules (DNA or RNA) in a stochastic regime; that is to say that we have a discrete number of molecules in a fixed volume. This regime is found in experimental systems that have a small volume with a fixed count of each molecule present, such as the interior of a cell, protocell, or droplet. We can also apply this to experimental systems with a larger volume (such as a test tube) when the system is well mixed, as we can either simulate a fixed small volume with small molecular counts and extrapolate to the larger volume, or we can individually simulate the interactions between specific molecules and derive rate constants for a coarse-grained chemical reaction network model that can be simulated in the mass-action regime.

To discuss the modeling and simulation of the system, we begin by defining the components of the system, and what comprises a state of the system within the simulation.

Strands. Each DNA molecule to be simulated is represented by a *strand*. Our system then contains a set of strands Ψ^*, where each strand $s \in \Psi^*$ is defined by $s = (id, label, sequence)$. A strand's *id* uniquely identifies the strand within the system, while the *sequence* is the ordered list of nucleotides that compose the strand. The strand label will usually be ignored, but may be used to make a distinction between strands with identical sequences. For example, if one strand were to be labeled with a fluorophore, it would no longer be physically identical to another with the same sequence but no fluorophore. We define two strands as being *identical* if they have the same labels and sequences.

Complex Microstate. A *complex* is a set of strands connected by base pairing (secondary structure). We define the state of a complex by $c = (ST, \pi^*, BP)$, called the *complex microstate*. The components are a nonempty set of strands $ST \subseteq \Psi^*$, an ordering π^* on the strands ST, and a list of base pairings $BP = \{(i_j \cdot k_l) \mid$ base i on strand j is paired to base k on strand l, and $j \leq l$, with $i < k$ if $j = l\}$, where "strand l" refers to the strand occurring in position l in the ordering π^*. Further, not all base pairings are allowed: following Ref. [4], every complex must by definition be connected, hairpins must have loop lengths of at least three, and in this work only non-pseudoknotted secondary structures will be considered.

System Microstate. A system microstate represents the configuration of the strands in the volume we are simulating (the "box"). We define a *system* *microstate i* as a set of complex microstates, such that each strand in the system is in exactly one complex within the system.

2.2 Energy

The conformation of a nucleic acid strand at equilibrium can be predicted by a well-studied model, called the nearest neighbor energy model [15–17]. Recent work has extended this model to cover systems with multiple interacting nucleic acid strands [4]. The distribution of system microstates at equilibrium is a Boltzmann distribution, where the probability of observing a microstate *i* is given by

$$Pr(i) = \frac{1}{Q_{kin}} * e^{-\Delta G_{box}(i)/RT} \tag{1}$$

where $\Delta G_{box}(i)$ is the free energy of the system microstate *i*, and is the key quantity determined by these energy models. $Q_{kin} = \sum_i e^{-\Delta G_{box}(i)/RT}$ is the partition function of the system, R is the gas constant, and T is the temperature of the system in Kelvin.

Energy of a System Microstate. To treat the energy of the system microstate *i*, we break it down into components. The system consists of one or more complex microstates *c*, each with their own energy. Additionally, the system has an entropy that accounts for the possible spatial arrangements of complexes within the "box".

Let us first consider the entropy term. Our reference state, which by definition will have zero energy, is chosen to be the system microstate in which all strands are in separate complexes and have no base pairs formed. Therefore, our entropy term is in terms of the reduction of available positional states caused by having strands join together. Assuming that the solution is sufficiently dilute that boundary and crowding effects can be ignored (i.e. each complex's center of mass can be anywhere within the simulated volume V), then each complex contributes $RT \log \frac{V}{V_0}$ to the energy of the system, where V_0 is the reference volume[1] chosen to be consistent with existing thermodynamic models. If the system contains L_{tot} strands within a total of C complexes, and we define $\Delta G_{volume} = RT \log \frac{V}{V_0}$, then the contribution to the energy of the system microstate *i* from the translational entropy of the box, relative to the reference state, is simply $(L_{tot} - C) * \Delta G_{volume}$.

And thus in terms of $C, L_{tot}, \Delta G_{volume}$ and $\overline{\Delta G}(c)$ (the energy of complex microstate *c*, defined in the next section), we define $\Delta G_{box}(i)$, the energy of the system microstate *i*, as follows:

---

[1] We calculate V_0 as the volume in which we would have exactly one molecule at a standard concentration of 1 mol/L: $V_0 = 1/(N_a * 1 \text{ mol/L})$, where N_a is Avogadro's number, and thus V_0 is in liters. Similarly, we may wish to calculate V based on the concentration u in mol/L of a single strand such that the volume V is chosen such that exactly one molecule is present in that volume. In this case we have $V = \frac{1}{u*N_a}$ and the relative number of states in the box is then $\frac{V}{V_0} = \frac{N_a}{u*N_a} = \frac{1}{u}$.

$$\Delta G_{box}(i) = (L_{tot} - C) * \Delta G_{volume} + \sum_{c \in i} \overline{\Delta G}(c) \qquad (2)$$

The energy formulas derived here, suitable for our stochastic model, differ from those in [4] in two main ways: the lack of "symmetry terms", and the addition of the ΔG_{volume} term.

Energy of a Complex Microstate. We previously defined a complex microstate in terms of the list of base pairings present within it. However, the well-studied models are based upon nearest neighbor interactions between the nucleic acid bases. These interactions divide the secondary structure of the system into local components which we refer to as *loops*, shown in Fig. 1.

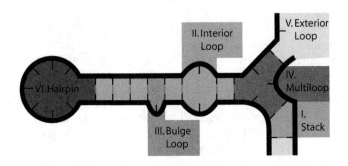

Fig. 1. Secondary structure divided into loops.

These loops can be broken down into different categories, and parameter tables and formulas for each category have been determined from experimental data [17]. Each loop l has an energy, $\Delta G(l)$, which can be retrieved from the appropriate parameter table for its category. Each complex also has an energy contribution associated with the entropic initiation cost [1] (e.g., rotational) of bringing two strands together, ΔG_{assoc}, whose total contribution is proportional to the number of strands L within the complex, as follows: $(L-1) * \Delta G_{assoc}$.

The energy of a complex microstate c is then the sum of these two types of contributions. We can also divide any free energy ΔG into the enthalpic and entropic components, ΔH and ΔS related by $\Delta G = \Delta H + T * \Delta S$. For a complex microstate, each loop can have both enthalpic and entropic components, but ΔG_{assoc} is usually assumed to be purely entropic [16]. This becomes important when determining the kinetic rates, in Sect. 2.3.

We use $\overline{\Delta G}(c)$ to refer to the energy of a complex microstate to be consistent with the nomenclature in [4], where $\overline{\Delta G}(c)$ refers to the energy of a complex when all strands within it are considered unique (as is the case in our system), and $\Delta G(c)$ is the energy of the complex, without assuming that all strands are unique (and thus it must account for rotational symmetries). In summary, the standard free energy of a complex microstate c, containing $L(c) = |ST(c)|$ strands, is

$$\overline{\Delta G}(c) = \left(\sum_{\text{loop } l \, \in c} \Delta G(l) \right) + (L(c) - 1)\Delta G_{assoc}$$

which can now be used in combination with Eq. 2 to compute $\Delta G_{box}(i)$.

It can also be convenient to write the system energy as a single sum over the complexes, rather than separating the complex microstate energies and the overall translational entropy terms. Using $L_{tot} = \sum_{c \in i} L(c)$, and $C = \sum_{c \in i} 1$, we obtain

$$\Delta G_{box}(i) = \sum_{c \in i} \left(\overline{\Delta G}(c) + (L(c) - 1) * \Delta G_{volume} \right) \overset{def}{=} \sum_{c \in i} \Delta G^*(c)$$

where

$$\Delta G^*(c) = \overline{\Delta G}(c) + (L(c) - 1) * \Delta G_{volume}$$

$$= \left(\sum_{\text{loop } l \in c} \Delta G(l) \right) + (L(c) - 1) * (\Delta G_{assoc} + \Delta G_{volume})$$

is the component of the total system energy that is associated with complex microstate c.

In Ref. [18], we have compared the Multistrand stochastic energy model to the NUPACK mass action model, showing that despite the lack of symmetry terms and the addition of ΔG_{volume} terms they nonetheless make identical equilibrium predictions.

2.3 Kinetics

Basics. Thermodynamic predictions have only limited use for some systems of interest, if the key information to be gathered is the reaction rates and not the equilibrium states. Many systems have well-defined ending states that can be found by thermodynamic prediction, but predicting whether it will reach the end state in a reasonable amount of time requires modeling the kinetics. Kinetic analysis can also help uncover poor sequence designs, such as those with alternate reactions leading to the same states, or kinetic traps which prevent an intended reaction from occurring quickly.

The kinetics are modeled as a continuous time Markov process over secondary structure space. System microstates i, j are considered adjacent if they differ by a single base pair (Fig. 2), and we choose the transition rates k_{ij} (the transition from state i to state j) and k_{ji} such that they obey detailed balance:

$$\frac{k_{ij}}{k_{ji}} = e^{-\frac{\Delta G_{box}(j) - \Delta G_{box}(i)}{RT}} \tag{3}$$

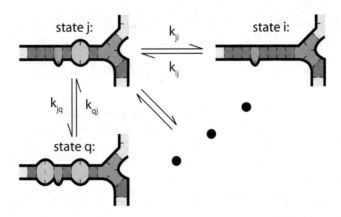

Fig. 2. System microstates i, q adjacent to current state j, with many others not shown.

This property ensures that given sufficient time we will arrive at the same equilibrium state distribution as the thermodynamic prediction (i.e., the Boltzmann distribution on system microstates, Eq. 1) but it does not fully define the kinetics as only the ratio $\frac{k_{ij}}{k_{ji}}$ is constrained. We discuss how to choose these transition rates in the following sections, but regardless of this choice we can still determine how the next state is chosen and the time at which that transition occurs.

Given that we are currently in state i, the next state m in a simulated trajectory is chosen randomly among the adjacent states j, weighted by the rate of transition to each.

$$Pr(m) = \frac{k_{im}}{\Sigma_j k_{ij}} \tag{4}$$

Similarly, the time taken to transition to the next state is chosen randomly from an exponential distribution with rate parameter λ, where λ is the total rate out of the current state, $\Sigma_j k_{ij}$.

$$Pr(\Delta t) = \lambda \exp(-\lambda \Delta t) \tag{5}$$

We will now classify transitions into two exclusive types: those that change the number of complexes present in the system, called *bimolecular transitions*, and those where changes are within a single complex, called *unimolecular transitions*. Note that this terminology is slightly different from the standard use of *bimolecular reactions* and *unimolecular reactions* in chemical reaction network theory: a bimolecular transition could be either a bimolecular reaction (two complexes coming together) or the corresponding unimolecular reaction (one complex dissociating into two).

Unimolecular Transitions. Because unimolecular transitions involve only a single complex, it is natural to define these transitions in terms of the complex microstate which changed, rather than the full system microstate. Like Fig. 2 implies, we define a complex microstate d as being adjacent to a complex microstate c if it differs by exactly one base pair. We call a transition from c

to d that adds a base pair a *creation move*, and a transition from c to d that removes a base pair a *deletion move*. The exclusion of pseudoknotted structures is not inherent in this definition of adjacent states, but rather arises from our disallowing pseudoknotted complex microstates.

The formal Markov chain for Multistrand simulations consists of transitions between system microstates i and j that differ by exactly one base pair, thus any unimolecular transition involves exactly one complex. Note that if i to j is a creation move, j to i must be a deletion move, and vice versa. Similarly, if there is no transition from i to j, there cannot be a transition from j to i, which implies that every unimolecular move in this system is reversible.

Bimolecular Transitions. A bimolecular transition from system microstate i to system microstate j is one where the single base pair difference between them leads to a differing number of complexes within each system microstate. This differing number of complexes could be due to a base pair joining two complexes in i to form a single complex in j, which we will call a *join move*. Conversely, the removal of this base pair from i could cause one complex in i to break into two complexes within j, which we will call a *break move*. Note that if i to j is a join move, then j to i must be a break move, and vice versa. As we saw before, this also implies that every bimolecular move is reversible. Again, while arbitrary bimolecular transitions are not inherently prevented from forming pseudoknots in this model, we implicitly prevent them by using only complex microstates that are not pseudoknotted.

Transition Rates. A key part of our model is the choice of *rate method*: the way we set the rates of a pair of reactions so that they obey detailed balance. There are several rate methods found in the literature [10,11,31] which have been used for kinetics models for single-stranded nucleic acids [7,31] with various energy models. We have implemented two of these simple rate methods which were previously used in single base pair elementary step kinetics models for single stranded systems.

In order to maintain consistency with known thermodynamic models, each pair of k_{ij} and k_{ji} must satisfy detailed balance and thus their ratio is determined by the thermodynamic model, but in principle each pair could be independently scaled by some arbitrary prefactor, perhaps chosen to optimize agreement with experimental results on nucleic acid kinetics. However, since the number of microstates is exponential, this leads to far more model parameters (the prefactors) than is warranted by available experimental data. For the time being, we limit ourselves to using only two scaling factors: k_{uni} for use with unimolecular transitions, and k_{bi} for bimolecular transitions.

Unimolecular Rate Models. The first rate model we will examine is the Kawasaki method [10]. This model has the property that both "downhill" (energetically favorable) and uphill transitions scale directly with the steepness of their slopes.

$$k_{ij} = k_{uni} * e^{-\frac{\Delta G_{box}(j) - \Delta G_{box}(i)}{2RT}} \tag{6}$$

$$k_{ji} = k_{uni} * e^{-\frac{\Delta G_{box}(i) - \Delta G_{box}(j)}{2RT}} \tag{7}$$

The second rate model under consideration is the Metropolis method [11]. In this model, all downhill moves occur at the same fixed rate, and only the uphill moves scale with the slope. This means that the maximum rate for any move is bounded, and in fact all downhill moves occur at this rate. This is in direct contrast to the Kawasaki method, where there is no bound on the maximum rate. For microstates i and j such that $\Delta G_{box}(i) \geq \Delta G_{box}(j)$:

$$k_{ij} = 1 * k_{uni} \qquad (8)$$

$$k_{ji} = k_{uni} * e^{-\frac{\Delta G_{box}(i) - \Delta G_{box}(j)}{RT}} \qquad (9)$$

Note that the value of k_{uni} that best fits experimental data is likely to be different for both models. Additionally, note that full calculation of $\Delta G_{box}(i)$ and $\Delta G_{box}(j)$ is not necessary in order to calculate the rates, because microstates i and j differ in exactly one pair of complex microstates ($c \in i, d \in j$) and by exactly three loop terms within those complex microstates.

Bimolecular Rate Model. When dealing with moves that join or break complexes, we must consider the choice of how to assign rates for each transition in a new light. In the particular situation of the join move, where two molecules in a stochastic regime collide and form a base pair, this rate is expected to be modeled by stochastic chemical kinetics.

Stochastic chemical kinetics theory [8] tells us that there should be a rate constant k such that the propensity of a particular bimolecular reaction between two species X and Y should be $k * \#X * \#Y/V$, where $\#X$ and $\#Y$ are the number of copies of X and Y in the volume V. Since our simulation considers each strand to be unique, $\#X = \#Y = 1$, and thus we see the propensity should scale as $1/V$. Recalling that $\Delta G_{volume} = RT \log \frac{V}{V_0} = RT \log \frac{1}{u}$, we see that we can obtain the $1/V$ scaling by letting the join rate be proportional to $e^{-\Delta G_{volume}/RT}$.

Thus we arrive at the following rate method, where the choice of the scalar term k_{bi} can be found by comparison to experiments measuring the hybridization rate of oligonucleotides [23], and where without loss of generality the transition from microstate i to microstate j is a join move while the transition from microstate j to microstate i is a break move:

$$k_{ij} = k_{bi} * e^{\frac{-\Delta G_{volume}}{RT}} = k_{bi} * \frac{V_0}{V} = k_{bi} * u \qquad (10)$$

$$k_{ji} = k_{bi} * e^{-\frac{\Delta G_{box}(i) - \Delta G_{box}(j) + \Delta G_{volume}}{RT}} \overset{def}{=} k_{bi} * e^{-\frac{\Delta G_{loops}(i,j) - \Delta G_{assoc}}{RT}} \qquad (11)$$

The latter simplification derives from the observation that, as in the bimolecular case, the system microstates i and j differ by exactly three loop terms in their complex microstates. However, they also differ in the total number of complexes within each system microstate, such that if i to j is a join move, $\Delta G_{box}(i) - \Delta G_{box}(j) = \Delta G_{loops}(i,j) - \Delta G_{volume} - \Delta G_{assoc}$, where $\Delta G_{loops}(i,j)$ represents the energy differences between i and j due to the three differing loop terms in the complex microstates.

This formulation is convenient for simulation, as the join rates are then independent of the resulting secondary structure. Note that an implication is that due to the rate being determined for **every** possible first base pair between two complexes, the overall rate for two complexes to bind (by a single base pair) is proportional roughly to the square of the number of exposed nucleotides (although possibly only a linear subset is likely to zipper up reliably), in addition to the $\frac{1}{V}$ dependence noted earlier.

3 The Simulator: Multistrand

Energy and kinetics models similar to these can been solved analytically; however, the standard master equation methods [24] scale with the size of the system's state space. For our DNA secondary structure state space, the size gets exponentially large as the strand length increases, so these methods become computationally prohibitive. One alternate method we can use is stochastic simulation [8], which has previously been done for single-stranded DNA and RNA folding (the **Kinfold** simulator [7]). Our stochastic simulation refines these methods for our particular energetics and kinetics models, which extends the simulator to handle systems with multiple strands and takes advantage of the localized energy model for DNA and RNA.

3.1 Data Structures

There are two main pieces that go into this new stochastic simulator. The first piece is the multiple data structures needed for the simulation: the *loop graph*, which represents the complex microstates contained within a system microstate (Fig. 3D); the *moves*, which represent transitions in our kinetics model (the single base pair changes in our structure that are the basic step in the Markov process); and the *move tree*, the container for moves that lets us efficiently store and organize them (Fig. 4).

Energy Model. Since the basic step for calculating the rate of a move involves the computation of a state's energy, we must be able to handle the energy model parameter set in a manner that simplifies this computation. Previous kinetic simulations (Kinfold) rely on the energy model we have described, though without the extension to multiple strand systems.

The energy model parameter set and calculations are implemented in a simple modular data structure that allows for both the energy computations at a local scale as we have previously mentioned, but also as a flexible subunit that can be extended to handle energy model parameter sets from different sources.

The Current State: Loop Structure. A complex microstate can be stored in many different ways, as shown in Fig. 3. While each of these has different advantages, we are going to focus on the loop representation, which allows the energy to be computed and stored in local components. One drawback is that the loop graph cannot represent pseudoknotted structures without introducing

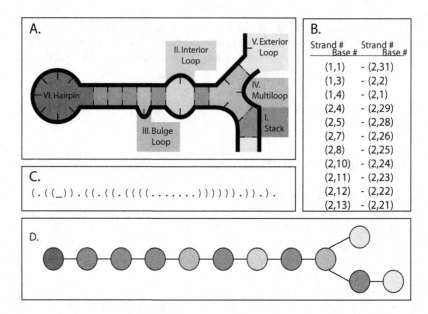

Fig. 3. Example secondary structure, with different representations: (A) Original loop diagram representation. (B) Base pair list representation. Each base pairing is represented by the indices of the bases involved. (C) Dot-paren representation, also called the flat representation. Each base is represented by either a period, representing an unpaired base, or by a parenthesis, representing a pairing with the base that has the (balanced) matching parenthesis. An underscore represents a break between multiple strands. (D) Loop graph representation. Each loop in the secondary structure is a single node in the graph, which contains the sequence information within the loop.

a loop type for pseudoknots (for which we may not know how to calculate the energy), and making the loop graph cyclic; however, since this work is primarily concerned with non-pseudoknotted structures this is only a minor point.

We use the loop graph representation for each complex within a system microstate, and organize those with a simple list. This gives us the advantage that the energy can be computed for each individual node in the graph, and since each move only affects either one or two nodes in the graph we will only have to recompute the energy for the affected nodes when performing a transition. While providing useful output of the current state then requires processing of the graph, it is a constant time operation if we store a flat representation which gets updated incrementally as each move is performed by the simulator.

Reachable States: Moves. When dealing with a flat representation or base pair list for a current state, we can simply store an available move as the indices of the bases involved in the move, as well as the rate at which the transition should occur. This approach is very straightforward to implement (as was done in the original Kinfold), and we can store all of the moves for the current state in a single global structure such as a list. However, when our current state is represented as a loop graph this simple representation can work, but does not

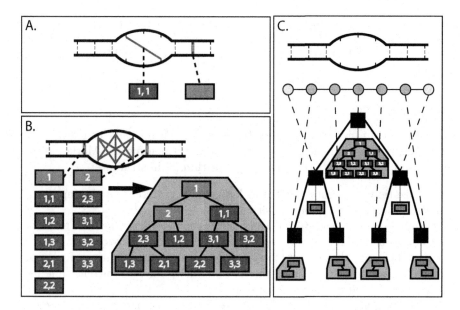

Fig. 4. (A) Creation moves (blue line) and deletion moves (red highlight) are represented here by rectangles. Either type of move is associated with a particular loop, and has indices to designate which bases within the loop are affected. (B) All possible moves which affect the interior loop in the center of the structure. These are then arranged into a tree (green area), which can be used to quickly choose a move. (C) Each loop in the loop graph then has a tree of moves that affect it, and we can arrange these into another tree (black boxes), each node of which is associated with a particular loop (dashed line) and thus a tree of moves (blue line). This resulting tree then contains all the moves available in the complex (Color figure online).

contain enough information to efficiently identify the loops affected by the move. Thus we elect to add enough complexity to how we store the moves so that we can quickly identify the affected nodes in our loop graph, which allows us to quickly identify the loops for which we need to recalculate the available moves.

We let each move contain a reference to the loop(s) it affects (Fig. 4A), as well as an index to the bases within the loop, such that we can uniquely identify the structural change that should be performed if this move is chosen. This reference allows us to quickly find the affected loop(s) once a move is chosen. We then collect all the moves which affect a particular loop and store them in a container associated with the loop (Fig. 4B). This allows us to quickly access all the moves associated with a loop whose structure is being modified by the current move. We should note that since deletion moves by nature affect the two loops adjacent to the base pair being deleted, they must necessarily show up in the available moves for either loop. This is handled by including a copy of the deletion move in each loop's moves, and halving the rate at which each occurs.

Finally, since this method of move storage is not a global structure, we add a final layer of complexity on top, so that we can easily access all the moves available from the current state without needing to traverse the loop graph. This is as simple as storing each loop's move container in a larger structure such as a list or a tree, which represents the entire complex's available moves as shown in Fig. 4C.

3.2   Algorithms

The second main piece of the simulator is the algorithms that control the individual steps of the simulator. The algorithm implementing the Markov process simulation closely follows the Gillespie algorithm [8] in structure:

1. Initialization: Generate the initial loop graph representing the input state, and compute the possible transitions.
2. Stochastic Step: Generate random numbers to determine the next transition, as well as the time interval elapsed before the transition occurs.
3. Update: Change the current loop graph to reflect the chosen move. Recompute the available transitions from the new state. Update the current time using the time interval in the previous step.
4. Check Stopping Conditions: check if we are at some predetermined stopping condition (such as a maximum amount of simulated time) and stop if it is met. Otherwise, go back to step 2.

The striking difference between this structure and the Gillespie algorithm is the necessity of recomputing the possible transitions from the current state at every step, and the complexity of that recalculation. Since we are dealing with an exponential state space we have no hope of storing all possible transitions between any possible pair of states, and instead must look at the transitions that occur only around the current state.

4   Analysis Case Studies

We have now presented the models and algorithms that form the continuous time Markov process simulator. Now we move on to discuss the most important part of the simulator from a user's perspective: the huge volume of data produced by the simulation, and methods for processing that data into useful information for analyzing the simulated system.

How much data are we talking about here? We would expect an average of $O(N)$ moves per time unit simulated, where N is the total length over all strands in the system. This doesn't tell us much about the actual amount of data, only that we expect it to not change drastically for different size input systems. In practice this amount can be quite large, even for simple systems: for a simple 25 base hairpin sequence, it takes $\sim$4,000,000 Markov steps to simulate 1 s of real time. For an even larger system, such as a four-way branch migration system with 108 total bases, simulating 1 s of real time takes $\sim$14,000,000 Markov steps.

What can we do with all the data produced by the simulator? A key insight is that most of this Markov step data is not needed if the measurement of

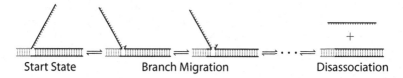

Fig. 5. Three-way branch migration system. The toehold region is in green and has sequence GTGGGT, and the branch migration region is black and has sequence ACCGCACCACGTGGGTGTCG. Both sequences are for the substrate strand.

interest is for a particular pathway, such as that shown in Fig. 5. In this system, one quantity of interest is how quickly the system reaches the completed branch migration state from the starting state. To measure this quantity we do not need to examine every Markov step as it is being made, but rather need to be able to record when we have reached the *stop state*. A stop state is typically defined by a *macrostate*, a collection of system microstates which meet some common criteria. For example, in the three way branch migration system, we might say that all system microstates which have the incumbent strand in a separate complex is the stop state of interest, as this corresponds to all the possible ways in which we could have had the incumbent strand dissociated at the end of the branch migration. For more on the definition of macrostates, see Ref. [18].

We now define the *first passage time mode* of simulation within Multistrand: given a starting system microstate and a set of stop states, it performs the simulation algorithm as given in Sect. 3.2 and records the time at which it reaches any of the stop states, as well as which one was reached. This produces a single piece of data for each trajectory simulated, which is a rather striking difference when compared to the raw number of microstates observed in a trajectory.

Let us now look at a simple three-way branch migration system in Fig. 5 and how it is to be simulated using first passage time mode. We start the system as shown, and use two different stop states: the *complete* stop condition where the incumbent strand has dissociated (as shown in the figure), and the *failed* stop condition where the invading strand has dissociated without completing the branch migration. Both of these are done using a macrostate describing a strand dissociation, which makes it very efficient to check the stop states. Note that we include the invading strand dissociating as a stop state so that if it occurs (which should be very rarely for long toehold lengths), we can find out easily without waiting until the maximum simulation time or until the strands reassociate and complete the branch migration.

The following table (Table 1) shows five trajectories' worth of data from first passage time mode on the example system. Note that we have included a third piece of data for each trajectory, which is the pseudorandom number generator seed used to simulate that trajectory. This allows us to produce the exact same trajectory again using a different simulation mode, stop states or other output conditions. For example, we might wish to run the fifth trajectory in the table again using trajectory mode, to see why it took longer than the others, or run the first trajectory to see what kinetic pathway it took to reach the failed stop condition.

Table 1. First passage time data for the example three-way branch migration system. Stop conditions are either "complete", indicating the branch migration completed successfully, or "failed", indicating the strands fell apart before the branch migration could complete.

Random number seed	Completion time	Stop condition
0x790e400d	$3.7 * 10^{-3}$	Failed
0x38188213	$3.8 * 10^{-3}$	Complete
0x47607ebf	$2.1 * 10^{-3}$	Complete
0x02efe7fa	$2.8 * 10^{-3}$	Complete
0x7c590233	$6.7 * 10^{-3}$	Complete

Let's now look at a much larger data set for first passage time mode. Here we again use the three-way branch migration system shown in Fig. 5, but with a ten base toehold region with sequence GTGGGTAGGT on the substrate strand in order to minimize the number of trajectories that reach the failed stop condition. We run 1000 trajectories, using a maximum simulation time of 1s, though no trajectory actually used that much as we shall shortly see.

Instead of listing all the trajectories in a table, we graph the first passage time data for the complete stop condition in two different ways: first (Fig. 6a) we make a histogram of the distribution of first passage times for the data set, and second (Fig. 6b) we graph the percentage of trajectories in our sample that have reached the complete stop condition as a function of the simulation time.

While there are many ways to analyze these figures, we note two particular observations. Firstly, the histogram of the first passage time distribution looks suspiciously like an exponential distribution, possibly with a short delay. This is not always typical, but the shape of this histogram can be very helpful in inferring

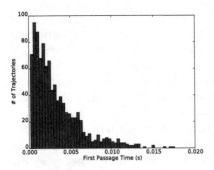

(a) Histogram of first passage times

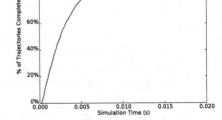

(b) Percent completion by simulation time

Fig. 6. First passage time data for the three-way branch migration system with ten base toehold. 1000 trajectories were simulated and all of them ended with the complete stop condition.

how we might wish to model our system based on the simulation data; e.g., for this system, we might decide that this three-way branch migration process is roughly exponential (with some fitted rate parameter) and so we could model it as a one-step unimolecular process.

The second observation is that while the percentage completion graph looks very similar to an experimental fluorescence microscopy curve, they should **NOT** be assumed to be directly comparable. The main pitfall is found when comparing fluorescence curves from systems where the reactions are bimolecular: in these the concentration of the relevant molecules are changing over time, but in our stochastic simulation the bimolecular steps are at a fixed volume/concentration (reflected in the ΔG_{volume} energy term) and data is aggregated over many trajectories.

5 Conclusions

The Multistrand simulator provides a powerful platform for exploring the behaviors of molecular machines created using dynamic DNA nanotechnology. In addition to the first passage time mode described above, alternative simulation modes have been implemented to provide differing levels of detail for analysis [18]: trajectory mode provides the full elementary step trajectory, which could be used to make a movie; transition mode collects statistics on when the simulation enters and exits specified macrostates; and first step mode runs simulations starting from an initial collision, which provides an efficient method for analyzing reactions in dilute solutions. The core simulation algorithms are implemented in C++, while a flexible user interface is available from within Python. The Multistrand package can be downloaded from http://www.multistrand.org.

At this time, Multistrand is best used to explore semi-quantitative sequence-dependent phenomena, such as assessing relative sequence design quality, because kinetic predictions are not expected to be in quantitative agreement with experimental measurements. While the secondary structure energy landscape used by Multistrand agrees with established thermodynamic models such as NUPACK [4], the simple methods used to set the relative rates of different types of elementary moves (Metropolis and Kawasaki dynamics) are not flexible enough to simultaneously accurately match the widely varying rates of fundamental processes such as zipping, fraying, breathing, three-way branch migration, and four-way branch migration. This is an important area for future work.

Acknowledgements. We are greatly indebted to years of insights, suggestions, and feedback from Niles Pierce, Robert Dirks, Justin Bois, and Victor Beck, especially their contributions to the formulation of the energy model and the first step simulation mode. This work has been funded by National Science Foundation grants DMS-0506468, CCF-0832824, CCF-1213127, CCF-1317694, and the Gordon and Betty Moore Foundation through the Caltech Programmable Molecular Technology Initiative.

References

1. Crothers, D.M., Bloomfield, V.A., Tinoco Jr., I.: Nucleic Acids: Structures, Properties, and Functions. University Science Books, Sausalito (2000)
2. Chen, Y.-J., Dalchau, N., Srinivas, N., Phillips, A., Cardelli, L., Soloveichik, D., Seelig, G.: Programmable chemical controllers made from DNA. Nat. Nanotechnol. **8**(10), 755–762 (2013)
3. Chitsaz, H., Salari, R., Sahinalp, S.C., Backofen, R.: A partition function algorithm for interacting nucleic acid strands. Bioinformatics **25**(12), i365–i373 (2009)
4. Dirks, R.M., Bois, J.S., Schaeffer, J.M., Winfree, E., Pierce, N.A.: Thermodynamic analysis of interacting nucleic acid strands. SIAM Rev. **49**(1), 65–88 (2007)
5. Dirks, R.M., Lin, M., Winfree, E., Pierce, N.A.: Paradigms for computational nucleic acid design. Nucleic Acids Res. **32**(4), 1392–1403 (2004)
6. Dirks, R.M., Pierce, N.A.: Triggered amplification by hybridization chain reaction. Proc. Natl. Acad. Sci. U. S. A. **101**(43), 15275–15278 (2004)
7. Flamm, C., Fontana, W., Hofacker, I.L., Schuster, P.: RNA folding at elementary step resolution. RNA **6**, 325–338 (2000)
8. Gillespie, D.T.: Exact stochastic simulation of coupled chemical reactions. J. Phys. Chem. **81**(25), 2340–2361 (1977)
9. Hongzhou, G., Chao, J., Xiao, S.-J., Seeman, N.C.: A proximity-based programmable DNA nanoscale assembly line. Nature **465**(7295), 202–205 (2010)
10. Kawasaki, K.: Diffusion constants near the critical point for time-dependent Ising models. Phys. Rev. **145**, 224–230 (1966)
11. Metropolis, N., Rosenbluth, A.W., Rosenbluth, M.N., Teller, A.H., Teller, E.: Equation of state calculations by fast computing machines. J. Chem. Phys. **21**, 1087–1092 (1953)
12. Muscat, R.A., Bath, J., Turberfield, A.J.: A programmable molecular robot. Nano Lett. **11**(3), 982–987 (2011)
13. Omabegho, T., Sha, R., Seeman, N.C.: A bipedal DNA Brownian motor with coordinated legs. Science **324**(5923), 67–71 (2009)
14. Qian, L., Winfree, E.: Scaling up digital circuit computation with DNA strand displacement cascades. Science **332**(6034), 1196–1201 (2011)
15. SantaLucia, J.: A unified view of polymer, dumbbell, and oligonucleotide DNA nearest-neighbor thermodynamics. Proc. Natl. Acad. Sci. **95**(4), 1460–1465 (1998)
16. SantaLucia, J., Allawi, H.T., Seneviratne, P.A.: Improved nearest-neighbor parameters for predicting DNA duplex stability. Biochemistry **35**(11), 3555–3562 (1996)
17. SantaLucia, J., Hicks, D.: The thermodynamics of DNA structural motifs. Ann. Rev. Biophys. Biomol. Struct. **33**(1), 415–440 (2004)
18. Schaeffer, J.M.: Stochastic simulation of the kinetics of multiple interacting nucleic acid strands. PhD thesis, California Institute of Technology (2013)
19. Seelig, G., Soloveichik, D., Zhang, D.Y., Winfree, E.: Enzyme-free nucleic acid logic circuits. Science **314**(5805), 1585–1588 (2006)
20. Shin, J.-S., Pierce, N.A.: A synthetic DNA walker for molecular transport. J. Am. Chem. Soc. **126**(35), 10834–10835 (2004)
21. Turberfield, A.J., Mitchell, J.C., Yurke, B., Mills Jr., A.P., Blakey, M.I., Simmel, F.C.: DNA fuel for free-running nanomachines. Phys. Rev. Lett. **90**(11), 118102 (2003)
22. Venkataraman, S., Dirks, R.M., Rothemund, P.W., Winfree, E., Pierce, N.A.: An autonomous polymerization motor powered by DNA hybridization. Nat. Nanotechnol. **2**(8), 490–494 (2007)

23. Wetmur, J.G.: Hybridization and renaturation kinetics of nucleic acids. Ann. Rev. Biophys. Bioeng. **5**(1), 337–361 (1976)
24. Wilkinson, D.J.: Stochastic dynamical systems. In: Stumpf, M.P., Balding, D.J., Girolami, M. (eds.) Handbook of Statistical Systems Biology, pp. 359–375. Wiley, New York (2011)
25. Xayaphoummine, A., Bucher, T., Isambert, H.: Kinefold web server for RNA/DNA folding path and structure prediction including pseudoknots and knots. Nucleic Acids Res. **33**(suppl 2), W605–W610 (2005)
26. Yin, P., Choi, H.M., Calvert, C.R., Pierce, N.A.: Programming biomolecular self-assembly pathways. Nature **451**(7176), 318–322 (2008)
27. Yurke, B., Turberfield, A.J., Mills, A.P., Simmel, F.C., Neumann, J.L.: A DNA-fuelled molecular machine made of DNA. Nature **406**(6796), 605–608 (2000)
28. Zadeh, J.N., Steenberg, C.D., Bois, J.S., Wolfe, B.R., Pierce, M.B., Khan, A.R., Dirks, R.M., Pierce, N.A.: NUPACK: analysis and design of nucleic acid systems. J. Comput. Chem. **32**(1), 170–173 (2011)
29. Zhang, D.Y., Seelig, G.: Dynamic DNA nanotechnology using strand-displacement reactions. Nature Chem. **3**(2), 103–113 (2011)
30. Zhang, D.Y., Turberfield, A.J., Yurke, B., Winfree, E.: Engineering entropy-driven reactions and networks catalyzed by DNA. Science **318**(5853), 1121–1125 (2007)
31. Zhang, W., Chen, S.-J.: RNA hairpin-folding kinetics. Proc. Natl. Acad. Sci. **99**(4), 1931–1936 (2002)

Author Index

Printed in the United States
By Bookmasters